SpringerBriefs in Computer Science

Series editors

Stan Zdonik, Brown University, Providence, Rhode Island, USA
Shashi Shekhar, University of Minnesota, Minneapolis, Minnesota, USA
Xindong Wu, University of Vermont, Burlington, Vermont, USA
Lakhmi C. Jain, University of South Australia, Adelaide, South Australia, Australia
David Padua, University of Illinois Urbana-Champaign, Urbana, Illinois, USA
Xuemin (Sherman) Shen, University of Waterloo, Waterloo, Ontario, Canada
Borko Furht, Florida Atlantic University, Boca raton, Florida, USA
V.S. Subrahmanian, University of Maryland, College Park, Maryland, USA
Martial Hebert, Carnegie Mellon University, Pittsburgh, Pennsylvania, USA
Katsushi Ikeuchi, University of Tokyo, Tokyo, Japan
Bruno Siciliano, Universita' di Napoli Federico II, Napoli, Italy
Sushil Jajodia, George Mason University, Fairfax, Virginia, USA
Newton Lee, Newton Lee Laboratories, LLC, Tujunga, California, USA

SpringerBriefs present concise summaries of cutting-edge research and practical applications across a wide spectrum of fields. Featuring compact volumes of 50 to 125 pages, the series covers a range of content from professional to academic.

Typical topics might include:

- A timely report of state-of-the art analytical techniques
- A bridge between new research results, as published in journal articles, and a contextual literature review
- A snapshot of a hot or emerging topic
- An in-depth case study or clinical example
- A presentation of core concepts that students must understand in order to make independent contributions

Briefs allow authors to present their ideas and readers to absorb them with minimal time investment. Briefs will be published as part of Springer's eBook collection, with millions of users worldwide. In addition, Briefs will be available for individual print and electronic purchase. Briefs are characterized by fast, global electronic dissemination, standard publishing contracts, easy-to-use manuscript preparation and formatting guidelines, and expedited production schedules. We aim for publication 8–12 weeks after acceptance. Both solicited and unsolicited manuscripts are considered for publication in this series.

More information about this series at http://www.springer.com/series/10028

Silvio Giancola • Matteo Valenti • Remo Sala

A Survey on 3D Cameras: Metrological Comparison of Time-of-Flight, Structured-Light and Active Stereoscopy Technologies

Silvio Giancola
Visual Computing Center
King Abdullah University of Science
Thuwal, Saudi Arabia

Matteo Valenti
Mechanical Engineering Department
Polytechnic University of Milan
Milan, Italy

Remo Sala
Polytechnic University of Milan
Milan, Italy

ISSN 2191-5768   ISSN 2191-5776   (electronic)
SpringerBriefs in Computer Science
ISBN 978-3-319-91760-3   ISBN 978-3-319-91761-0   (eBook)
https://doi.org/10.1007/978-3-319-91761-0

Library of Congress Control Number: 2018942612

Printed on acid-free paper

This Springer imprint is published by the registered company Springer International Publishing AG part of Springer Nature.
The registered company address is: Gewerbestrasse 11, 6330 Cham, Switzerland

Preface

Metrology, from the Greek *Metro-logos*, is the logic (*-Logos*) ruling the study of measurement (*Metro-*), which has been active for more than two centuries. Research in metrology focuses on establishing a common knowledge of physical quantities. The Bureau International des Poids et Mesures (BIPM) enforces a universal way to define and use such physical quantities with the International System (SI). Regularly, the BIPM updates the rules that dictate how to perform measurements (BIPM et al. 2008). Through the Guide to the Expression of Uncertainty in Measurement (GUM), they provide the methodology and the vocabulary to assess the uncertainty of a measurement, as well as the performances of an instrument.

In this work, we attempt to apply the rigorous methodology of the GUM within the field of computer vision. We deliver our manuscript as a practical user manual for three-dimensional (3D) cameras. We provide the reader with our experience in testing, calibrating and using 3D cameras. We propose a deep-enough understanding of the underlying technology as well as a comparative study of the commercially available 3D cameras. We hope to provide enough insight in our manuscript to help identifying the optimal device or technology for a given application.

This manuscript is the fruit of research focusing on understanding and evaluating non-contact measurements based on computer vision technology. While most of the experiments were realized in the Mechanical Engineering Department of Politecnico di Milano in Italy, part of them were realized in the Visual Computing Center (VCC) of King Abdullah University of Science and Technology (KAUST) in Saudi Arabia. Such enterprise would not have been possible without the contribution of several people: We thank Alessandro Basso, Mario Galimberti, Giacomo Mainetti and Ambra Vandone for their introduction of metrology to computer vision; Andrea Corti, Nicolo Silvestri and Alessandro Guglielmina for their contribution in the metrological analysis of the depth cameras; PierPaolo Ruttico and Carlo Beltracchi for their valuable contribution to the tests on Intel devices; Moetaz Abbas for its consultancy and the analysis of Time-of-Flight (TOF) signal; Matteo Matteucci and Per-Erik Forssen for the valuable technical feedback on 3D computer vision; Matteo Scaccabarozzi, Marco Tarabini and Alfredo Cigada for sharing their knowledge in metrology; Bernard Ghanem and Jean Lahoud for sharing their knowledge in computer vision. Also, we thank the fantastic and exciting computer vision and metrology communities who provide us valuable feedbacks.

Thuwal, Saudi Arabia   Silvio Giancola
Milano, Italy   Matteo Valenti
Milano, Italy   Remo Sala
November 2017

Contents

Acronyms

2D	Two-dimension
3D	Three-dimension
ASIC	Application-Specific Integrated Circuit
BIPM	Bureau International des Poids et Mesures
CCD	Charge-Coupled Device
CMOS	Complementary Metal-Oxide-Semiconductor
CT	Computer Tomography
CW	Continuous-Wave
FOV	Field of View
GAPD	Geiger-mode Avalanche Photo Diode
GPU	Graphics Processing Unit
GUM	Guide to the Expression of Uncertainty in Measurement
ICP	Iterative Closest Point
IR	Infra-Red
LiDAR	Light Detection And Ranging
NIR	Near Infra-Red
PCL	Point Cloud Library
RADAR	Radio Detection and Ranging
RANSAC	RANdom SAmple Consensus
SDK	Software Development Kit
SfM	Structure-from-Motion
SI	International System
SNR	Signal-to-Noise Ratio
SONAR	Sound Detection and Ranging
SPAD	Single-Photon Avalanche Diode
SRS	Spatial Reference System
SVD	Singular Value Decomposition
TOF	Time-of-Flight
UV	Ultra-Violet

Chapter 1
Introduction

Studies in computer vision attempts to understand a given scene using visual information. From a hardware perspective, vision systems are transducers that measure the light intensity. They usually produce images or videos but can also generate point clouds or meshes. From a software perspective, vision algorithms attempt to mimic the natural human process. They usually focus on detecting and tracking objects or reconstructing geometrical shapes.

In its simplest form, Two-dimension (2D) computer vision processes images or videos acquired from a camera. Cameras are projective devices that capture the visual contents of the surrounding environment. They measure the color information, seen from a fixed point of view. Traditional cameras provide solely flat images and lack of geometrical knowledge. The issue of depth estimation is tackled in Three-dimension (3D) computer vision by carefully coupling hardware and software. Among those, 3D cameras capture range maps aside the color images. They recently gained interest among the computer vision community, thanks to their democratization, their price drop and their wide range of application.

3D cameras and 3D devices are commonly used in numerous applications. For topographic engineering, laser scanners are commonly used for the reconstruction of large structures such as bridges, roads or buildings. For cultural heritage documentation, laser scanner devices and Structure-from-Motion (SfM) techniques enable the reconstruction of archaeological finds or ancient objects. In radiology, 3D devices such as Computer Tomography (CT) are used to see within the human body. In physical rehabilitation, 3D vision systems are used to track and analyze human motion. Similarly in movies, 3D vision systems are used to track actors and animate digital characters. For video game entertainment, 3D cameras enhance the player interface within a game. In robotics, 3D vision systems are used to localize autonomous agents within a map of the surrounding environment. It also provides the sense of perception to detect and recognize objects. For the manufacturing

© The Author(s), under exclusive licence to Springer International Publishing AG, part of Springer Nature 2018
S. Giancola et al., *A Survey on 3D Cameras: Metrological Comparison of Time-of-Flight, Structured-Light and Active Stereoscopy Technologies*, SpringerBriefs in Computer Science, https://doi.org/10.1007/978-3-319-91761-0_1

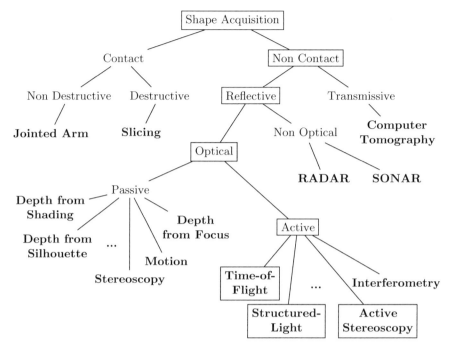

Fig. 1.1 Taxonomy for **3**D reconstruction techniques. In this book we are focusing on Time-of-Flight, Structured-Light and Active Stereo

industry, reliable **3**D vision systems are used in autonomous assembly line to detect and localize objects in space.

3D vision devices can be considered as a tool to acquire shape. **3**D shape acquisition covers a field of study wider than computer vision. It exists numerous systems based on various technologies. An overview is given in Fig. 1.1.

First of all, **3**D shape acquisition can be split between *Contact* and *Non Contact* techniques. *Contact* techniques can be *destructive*, such as **Slicing**, that reduces the dimension of the analysis by sectioning an object into **2**D shapes successively assembled together. It can also be *non destructive*, such as **Jointed arms**, that slowly but accurately probes **3**D points. *Non Contact* techniques usually measures areas instead of single spots on a target. They avoid any physical contact with the object to measure, hence remove any loading effects and avoid damaging the object to measure.

Non Contact techniques can be divided into *Reflective* and *Transmissive* ones, the former using the reflection of a signal emitted from a body, the latter exploiting its transmission. For instance, **Computer Tomography** is a *Transmissive* technique that uses X-rays signals taken from different poses to identify changes in density within a body. Alternatively, *Reflective* techniques focus on analyzing signals reflection. *Non Optical* techniques focuses on wavelength that are not comprise within the visible or the infrared spectrum. **Sound Detection and Ranging (SONAR)**, that uses sound signals and **Radio Detection and Ranging (RADAR)**, that uses radio signals are examples of *Non Optical* techniques that estimate range maps on long distances by estimating the time the signals run through its environment.

Optical techniques exploits the visible (400–800 nm) and the Infra-Red (IR) (0.8–1000μm) wavelengths to get information from a scene or a object. While color is commonly used since it is mimicking the human vision system, IR wavelengths carry out temperature information and is usually more robust to ambient light. *Optical* techniques for shape acquisition can be furthermore divided into *Passive* and *Active* methods. *Passive* methods use the reflection of natural light on a given target to measure its shape. **Stereoscopy** looks for homogeneous features from multiple cameras to reconstruct a 3D shape, using traingulation and epipolar geometry theory. Similarly, **Motion** exploits a single camera that moves around the object. **Shape from Silhouette** and **Shape from Shading** allow direct and simple shape measurement based on the edges and shading theory. **Depth of Field** uses the focus information of the pixels given a sensor focal length to estimate its range.

Active methods enhance shape acquisition by using an external lighting source that provides additional information. Similar than SONAR and RADAR, **Time-of-Flight** systems are based on the Light Detection And Ranging (LiDAR) principle. Time-of-Flight systems estimate the depth by sending lighting signals on the scene and measuring the time the light signal goes back-and-forth. **Structured-Light** devices project a laser pattern to the target and estimate the depth by triangulation. Sub-millimeter accuracy can be reached with laser blade triangulation, but only estimate the depth along a single dimension. To cope with depth maps, Structured-Light cameras project a 2D codified patterns to perform triangulation with. **Active Stereoscopy** principle is similar to the passive one, but looks for artificially projected features. In contrast with Structured-Light, the projected pattern is not codified and only serves as additional features to triangulate with. Finally, **Interferometry** projects series of fringes such as *Moire*'s to estimate shapes. Such method requires an iterative spatial refinement in the projected pattern hence is not suitable for depth map estimation from a single frame.

In this book, we focus the attention on active 3D cameras. 3D cameras extract range maps, providing depth information aside the color one. Recent 3D cameras are based on Time-of-Flight, Structured-Light and Active Stereoscopy technologies. We organize the manuscript as following: In Chap. 2, we present the camera model as well as the Structured-Light, Active Stereoscopy and Time-of-Flight (TOF) technologies for 3D shape acquisition. In Chap. 3, we provide an overview of the 3D cameras commercially available. In Chaps. 4–6, we provide an extended metrological analysis for the most promising 3D cameras based on the three aforementioned technologies, namely the Kinect V2, the Orbbec Astra S and the Intel RS400 generation.

Chapter 2
3D Shape Acquisition

3D cameras are matrix sensors that estimate depth and capture range maps. Similar to color cameras, they provide images of the surrounding environment as seen from a single point of view. Alongside the color information, Three-dimension (3D) cameras provides depth measurements by exploiting visual information. Different techniques exist to measure 3D shapes, mainly by triangulating keypoints from two point of view or by directly estimating the range (Fig. 2.1). In this section, we present the theoretical foundation behind common techniques used in 3D cameras. First, we introduce the linear camera model and show the non-linearity introduced by the optical lens (Sect. 2.1). Second, we provide the theoretical background for estimating depth through triangulation (Sect. 2.2) and Time-of-Flight (TOF) (Sect. 2.3). Last, we elaborate on the depth maps to point cloud transformation and on the color information integration.

2.1 Camera Model

The camera model is an inherent part of any 3D camera. A camera is a device that captures light information provided by an environment, transforms it in a processable physical quantity and visualizes its measurement map as seen from its point of view. A camera is considered as a *non-contact* (it does not physically interfere with the target) and *optical* (it takes advantages of light properties) device. Intrinsically, a camera is a *passive* optical device since it only measures the light provided by the surrounding environment. Nevertheless, for the 3D camera we are presenting, the depth measurement is using an artificial lighting system which makes it *active*. In the following, we recall the camera model developing its linear and non-linear models.

S. Giancola et al., *A Survey on 3D Cameras: Metrological Comparison of Time-of-Flight, Structured-Light and Active Stereoscopy Technologies*, SpringerBriefs in Computer Science, https://doi.org/10.1007/978-3-319-91761-0_2

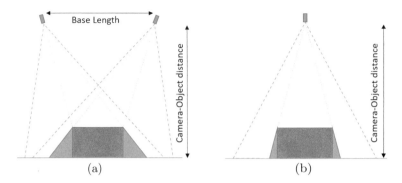

Fig. 2.1 Comparison between **3D** camera measurement principles. The **3D** shape of interest are the orange boxes, observed by the camera with field of view displayed in blue. Note the *penumbra* in gray. (**a**) Triangulation. (**b**) Direct depth estimation

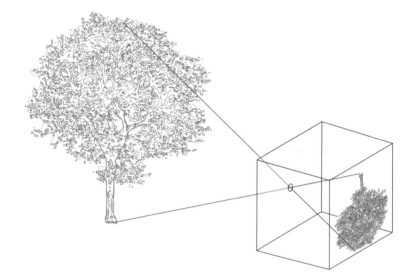

Fig. 2.2 *Pinhole camera* principle (Hartley and Zisserman 2003)

2.1.1   Linear Camera Model

A simplified representation of the camera system is referred as the *Pinhole Camera* presented in Hartley and Zisserman (2003), also related as the *camera obscura*, depicted in Fig. 2.2. The light emitted by the environment enters on a lightproof chamber through a pinhole. The small dimension of the hole, ideally a point, prevents the diffusion of the light, that travels in a straight line hitting the opposite side of the chamber. The *camera obscura* produces an overturned projection of the environment, as seen from the pinhole point of view.

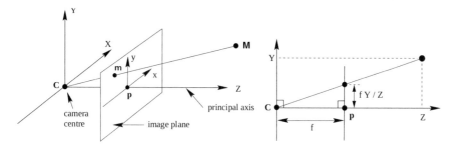

Fig. 2.3 *Left:* Frontal pinhole camera representation. *Right:* Trigonometrical representation

The frontal geometrical pinhole model illustrates the geometry of the phenomenon, as shown in Fig. 2.3. The *optical center* **C**, also referred as the *camera center*, represents the pinhole, formed by a lens, from where the light enters inside the device. The lens task is to focus the light crossing the pinhole on the plane called *image plane*. The **3D** environment is projected on the image plane. Note that this plane is represented in front of the camera center in order to avoid the overturned projection model, but physically, this plane is behind the lens. The *principal axis* corresponds to the **Z**-axis of the Spatial Reference System (SRS) positioned in the optical center **C**. The projection of the light rays that crosses the lens are sensed by a matrix of photo-diodes, usually rectangular, that define the **X**- and **Y**-axes, spaced from the focal distance f.

Equation (2.1) sets up the projection transformation of a **3D** point **M** with coordinates (X, Y, Z) in a Two-dimension (**2D**) point **m** with coordinates (x, y). Equation (2.2) introduces homogeneous coordinates for the projective development. The 3×4 matrix **P** is called *projection matrix* or *camera matrix*.

$$\mathbf{M} = \{X, Y, Z\}^T \rightarrow \mathbf{m} = \{x, y\}^T = \left\{ f\frac{X}{Z}, f\frac{Y}{Z} \right\}^T \tag{2.1}$$

$$\begin{pmatrix} fX/Z \\ fY/Z \\ 1 \end{pmatrix} = \frac{1}{Z} \begin{bmatrix} f & 0 & 0 & 0 \\ 0 & f & 0 & 0 \\ 0 & 0 & 1 & 0 \end{bmatrix} \cdot \begin{pmatrix} X \\ Y \\ Z \\ 1 \end{pmatrix} = \mathbf{P} \cdot \begin{pmatrix} X \\ Y \\ Z \\ 1 \end{pmatrix} \tag{2.2}$$

Sensors composed of a matrix of photo-diodes are used to capture visual light information on the image plane. They are transducers that exploit the photo-voltaic effect to generate an electric signal proportional to the number of photons that strikes them. Electrons migrate from the valence band to the conduction band, creating an electromotive force in function of the quantity of photons. Wavelengths for optical systems range from 100 nm (Ultra-Violet (UV)) to 1 mm (Infra-Red (IR)). Grabbing from a matrix sensor will sample the scene projection in digital images. Images are matrices of measured photons quantities, the which information is usually digitized in 8 to 24 bits. It exists two types of architecture for the photo-diodes matrices,

Charge-Coupled Device (CCD) and *Complementary Metal-Oxide-Semiconductor (CMOS)*.

The CCD sensors have a unique analog-to-digital converter and processing block that transduces all the photo-diode signals into digital values. Usually, such sensor are wider than CMOS. Since the matrix is only composed of photo-diodes, pixels can be larger thus capture more photons and produce better quality images. Also, using a single superior amplifier and converter produces more uniform images with less granularity. Nevertheless, due the singularity of the electrical conversion, rolling shutter is mandatory and grabbing images with high dynamic motion will not return consistent images. They generally present a higher cost of production and are usually used in photogrammetry.

The CMOS sensors incorporate a dedicated analog-to-digital converter and processing block for each photo-diode, usually of lower quality respect to the CCD one's. Typically, the sensor has a smaller scale factor with a lower photo-diode exposition area and the captured image is less uniform and noisier, since each photo-diode carries out its own conversion. Nevertheless, the independence of the pixel improves the modularity of those sensors, it is possible to acquire a portion of the sensor with a global shutter principle. They are typically used in the mobile market due to their lower cost and in industrial applications thanks to their modularity and fast acquisition rates.

Taking into account that the light on the image plane is measured through a rectangular matrix of pixels, it is possible to improve the projection model of the 3D point $\mathbf{M}$ on the matrix sensor SRS. A first consideration consists of a translation of the sensor SRS from the optical center $\mathbf{C}$ of the camera to a corner of the image. Supposing that the image plane is parallel to the $(\mathbf{C}, \mathbf{X}, \mathbf{Y})$ plane, the transformation consists in a translation c_x and c_y along $\mathbf{X}$- and $\mathbf{Y}$-axes, as shown in Fig. 2.4. The c_x and c_y values correspond to the optical center coordinates in pixel to the bottom-left corner of the sensor. Note that the knowledge of the pixel size of the sensor, usually around the μm, permits the conversion from pixels to SI units.

Equation (2.2) becomes Eqs. (2.3) and (2.4) introducing the *calibration matrix* $\mathbf{K}$. $\mathbf{M}$ is the homogeneous coordinate of a point in space and $\mathbf{m}$ its projection in the image plane. Z is the depth coordinate of the point $\mathbf{M}$ within the camera SRS.

Fig. 2.4 Optical center translation

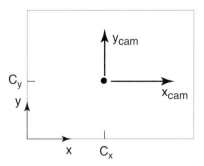

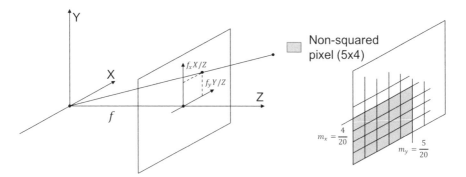

Fig. 2.5 Matrix sensor with rectangular pixels

$$\begin{pmatrix} fX/Z + c_x \\ fY/Z + c_y \\ 1 \end{pmatrix} = \frac{1}{Z} \begin{bmatrix} f & 0 & c_x & 0 \\ 0 & f & c_y & 0 \\ 0 & 0 & 1 & 0 \end{bmatrix} \begin{pmatrix} X \\ Y \\ Z \\ 1 \end{pmatrix} \tag{2.3}$$

$$\mathbf{m} = \frac{1}{Z} \mathbf{P} \cdot \mathbf{M} = \mathbf{K} \cdot [\mathbf{I}|0] \cdot \mathbf{M} \quad \text{with} \quad \mathbf{K} = \begin{bmatrix} f & 0 & c_x \\ 0 & f & c_y \\ 0 & 0 & 1 \end{bmatrix} \tag{2.4}$$

Note that the sensor pixels are not always squared. Since measurements on the image are usually provided in pixels, the horizontal and the vertical unit vectors are different and their composition becomes tricky. This effect is illustrated in Fig. 2.5; non squared dimension of the pixel results in a different number of pixel per length unit which leads to a different scale factor in the projection along the two directions of the plane. In order to take this phenomenon into account, the focal length f is split into two components $f_x = f \cdot m_x$ and $f_y = f \cdot m_y$, m_x and m_y being two factors that represent the number of pixels per length unit along the **X**- and **Y**-axes.

Successively, pixels do not always have rectangular shape with perpendicular sides. The *skew* parameter $s = f \cos(\alpha)$ introduces this correction, α being the angle between two sides of the pixel. Note that such parameter is usually considered as null ($\alpha = 90°$). The corrected calibration matrix (**K**) taking into account the non-squarely and the non-perpendicularity aspect of the pixels is presented on the *calibration matrix* in Eq. (2.5).

$$\mathbf{K} = \begin{bmatrix} f_x & s & c_x \\ 0 & f_y & c_y \\ 0 & 0 & 1 \end{bmatrix} \tag{2.5}$$

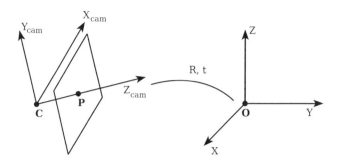

Fig. 2.6 Camera pose in the global SRS

In most cases, the camera is placed arbitrarily in an environment, an $SE(3)$ transformation defines the camera SRS, as shown in Fig. 2.6. The camera pose is composed of an orientation $\mathbf{R}$ and a translation $\mathbf{t} = -\mathbf{R} \cdot \mathbf{C}$, $\mathbf{C}$ being the coordinates of the camera in the global SRS. Equation (2.6) summarizes the projection operation that occurs in the camera, taking into account both intrinsic and extrinsic parameters.

$$\mathbf{m} = \frac{1}{Z}\mathbf{K}[\mathbf{R}|\mathbf{t}]\mathbf{M} \quad \text{with} \quad \mathbf{t} = -\mathbf{RC} \tag{2.6}$$

A total of 11 parameters define the linear model of the camera. In numerous applications, this model offers a good approximation of the reality. Nevertheless, it is not accurate enough when important distortion are present, due to the presence of a lens in the pinhole. Those distortions introduce a non-linearity in the model, presented in the next section.

2.1.2   Non-linear Camera Model

The presence of a lens that conveys the light rays on the sensor pixels creates non-linear distortions. It is known that smaller the focal length is, more apparent are those distortions. Fortunately, they can be modeled and estimated thanks to an opportune calibration. The non-linear model allow for the rectification of any images captured by the camera and permit coherent measurements on it.

Distortions are non-linear phenomena that occurs when the light crosses the lens of the camera, due to its non-perfect thinness, especially for short focal length. The effects of distortions are presented in Fig. 2.7, where a straight line on the calibration plate projects as a curve in the image instead of a line. After rectification of the image, taking into account the *radial* and *tangential* distortion models explained successively, the linear geometrical consistency is maintained.

The non-linear formula that corrects any radial and tangential distortion is presented in Eq. (2.7), where (x, y) are the coordinates of the pixels in the image, $(\hat{x}, \hat{y})$ are the corrected coordinates and (c_x, c_y) are the coordinates of the optical center.

Fig. 2.7 *Left:* Distorted image. *Right:* Rectified image

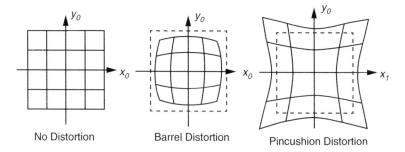

No Distortion   Barrel Distortion   Pincushion Distortion

Fig. 2.8 Illustration of *barrel* or *pincushion* distortions

$$\begin{cases} \hat{x} = c_x + (x - c_x)(k_1 r^2 + k_2 r^4 + \cdots) \\ \hat{y} = c_y + (y - c_y)(k_1 r^2 + k_2 r^4 + \cdots) \end{cases} \quad \text{with } r = (x - c_x)^2 + (y - c_y)^2$$

$$(2.7)$$

Globally, the further from the optical center a pixel is, the more correction it undergoes. Note that the $(k_1 r^2 + k_2 r^4 + \cdots)$ factor is a Taylor series where the parameters $k_{i \in \mathbb{N}^*}$ are usually shelved at $i = 3$. The distortions can be present in two modalities, in function of the signs of the radial parameters, *barrel* or *pincushion* ones (Fig. 2.8).

A better model of distortion has been introduced by Brown (1966), known as the *Plumb Bob* model. This model does not consider the sole radial distortions but introduces additional tangential distortions, attributed to an error in fabricating and mounting the lens. Figure 2.9 illustrates the differences between both distortions. The complete relation between rectified coordinates $(\hat{x}, \hat{y})$ and the original one (x, y) are given in Eq. (2.8), introducing two tangential distortion parameters p_1 and p_2.

$$\begin{cases} \hat{x} = (1 + k_1 r^2 + k_2 r^4 + k_3 r^6)x + 2p_1 xy + p_2(r^2 + 2x^2) \\ \hat{y} = (1 + k_1 r^2 + k_2 r^4 + k_3 r^6)y + 2p_2 xy + p_1(r^2 + 2y^2) \end{cases}$$

$$(2.8)$$

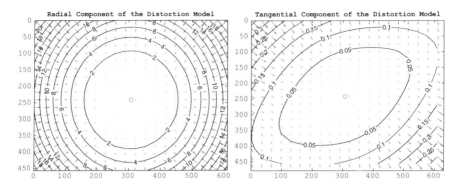

Fig. 2.9 Quantification of radial (left) and tangential (right) distortions

Altogether, a total of 16 parameters define the non-linear camera model:

- Five internal parameters, from the calibration matrix $\mathbf{K}$ (f_x, f_y, c_x, c_y, s),
- Six external parameters, from the position ($\mathbf{t}$) and orientation ($\mathbf{R}$) of such camera in a global SRS,
- Five distortion parameters, split into radials (k_1, k_2, k_3) and tangentials (p_1, p_2).

Those parameters are necessary to rectify the image acquired from the matrix sensor acquisition. Rectified images are required to realize reliable measurements on a scene. Mainetti (2011) provides an extensive evaluation of calibration processes for the determination of such parameters.

2.2   Depth by Triangulation

In order to extract 3D geometrical information, triangulation methods estimate the depth by observing the target from different prospective. Triangulation can occur in various fashion. *Passive Stereoscopy* uses two cameras to triangulate over homologous keypoints on the scene (Sect. 2.2.1). *Active Stereoscopy* uses two cameras and a light projector to triangulate with the two cameras over the artificial features provided by the projector (Sect. 2.2.4). *Structured-Light* uses one camera and a structured light projector and triangulates over the codified rays projected over the scene (Sect. 2.2.5).

2.2.1   Stereoscopy

Stereoscopy is inspired by the human brain capacity to estimate the depth of a target from images captured by the two eyes. Stereoscopy reconstructs depth by exploiting the disparity occurring between cameras frames that capture the same scene from different points of view.

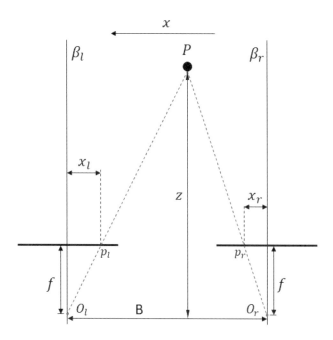

Fig. 2.10 Basic stereo system

Figure 2.10 shows a simplified model for the triangulation in stereoscopy, with cameras laterally translated from each others. In this illustration, the two cameras have parallel optical axes and observe a point **P** located a distance z. The optical projections $\mathbf{p}_l$ and $\mathbf{p}_r$ of **P** are shown on the left and right camera plane. Note the similarity between the triangles $O_l P O_r$ and $p_l P p_r$. Knowing the intrinsic and extrinsic parameters of the cameras, the depth z is inversely proportional to the horizontal parallax between projections $\mathbf{p}_l$ and $\mathbf{p}_r$, as shown in Eq. (2.9). Note that f, x_l, x_r are provided in pixel, hence the depth z has the same unit as the translation B. In literature $(x_r - x_l)$ is defined as the *disparity d* and the optical centers distance as the *baseline B*.

$$z = \frac{Bf}{(x_r - x_l)} \tag{2.9}$$

Considering Eq. (2.9), the derivative with respect to the disparity d is shown in Eq. (2.10) where an error in depth ∂z grows quadratically with the depth z. Note that the error in disparity ∂d depends on the algorithm; common strategies ensure subpixel values. Measurements on commercial techniques such as the Intel's R200 showed a RMS value for the disparity error of about 0.08 pixel, in static condition.

$$\partial z = \frac{z^2}{Bf} \partial d \tag{2.10}$$

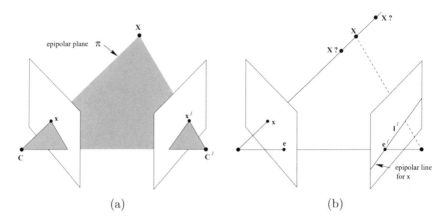

Fig. 2.11 Epipolar geometry (Hartley and Zisserman 2003)

2.2.2   *Epipolar Geometry*

The triangulation process used in stereoscopy is based on the epipolar geometry. The epipolar geometry describes the geometrical relationships between two different perspective view of a same 3D scene.

Consider a more general setup with two camera system as sketched in Fig. 2.11a. The two camera centers are defined by their centers $\mathbf{C}$ and $\mathbf{C}'$ and the 3D point $\mathbf{X}$ projects in both camera plane in $\mathbf{x}$ and $\mathbf{x}'$, respectively. The *epipolar plane* for $\mathbf{X}$ is defined by the two camera centers $\mathbf{C}$, $\mathbf{C}'$ and the point $\mathbf{X}$ itself. In Fig. 2.11b, the epipolar plane intersect the second camera plane in the *epipolar lines* $\mathbf{l}'$. Also, the line between the two camera centers is called the *baseline* and intersects the image plane at the *epipole* $\mathbf{e}$ and $\mathbf{e}'$. In other words, the *epipoles* are the projections of a camera center on the other camera image plane.

Given the sole projection $\mathbf{x}$ on the camera defined with its center $\mathbf{C}$, the depth of the 3D point $\mathbf{X}$ is ambiguous. By knowing the relative pose of the second camera defined by its center $\mathbf{C}'$, the identified epipolar plane intersects the second camera image plane in the epipolar line. As a result, finding the projection $\mathbf{x}'$ of $\mathbf{X}$ in the second camera along the epipolar line $\mathbf{l}'$ will solve the depth ambiguity.

Algebraically, the *fundamental matrix* $\mathbf{F}$ is defined as a rank 2 homogeneous matrix with 7 degree of a freedom that solve Eq. (2.11), for any homologous points $\mathbf{x}$ and $\mathbf{x}'$. From a computational point of view, such matrix can be retrieved by an *a-priori* knowledge of at least eight corresponding points between frames or through a calibration process.

$$\mathbf{x}'^{T}\mathbf{F}\mathbf{x} = 0 \qquad\qquad (2.11)$$

Equation (2.12) holds since the epipoles $\mathbf{e}$, $\mathbf{e}'$ represent the projections of each camera's center $\mathbf{C}$, $\mathbf{C}'$ on the other one's frame, with $\mathbf{P}$ and $\mathbf{P}'$ the cameras projection matrices.

$$\mathbf{e} = \mathbf{P}\mathbf{C}' \qquad\qquad \mathbf{e}' = \mathbf{P}'\mathbf{C} \qquad\qquad (2.12)$$

The *fundamental matrix* $\mathbf{F}$ is obtained as function of the cameras projection matrices in Eq. (2.13), where $\mathbf{P}^\dagger$ is the pseudo-inverse and the $[\mathbf{e}']_\times$ the skew symmetric matrix from the vector $\mathbf{e}'$.

$$\mathbf{F} = [\mathbf{e}']_\times \mathbf{P}'\mathbf{P}^\dagger = \begin{bmatrix} 0 & \mathbf{e}'_1 & -\mathbf{e}'_2 \\ -\mathbf{e}'_1 & 0 & \mathbf{e}'_3 \\ \mathbf{e}'_2 & -\mathbf{e}'_3 & 0 \end{bmatrix} \mathbf{P}'\mathbf{P}^\dagger \qquad\qquad (2.13)$$

The epipolar lines $\mathbf{l}'$ and $\mathbf{l}$ related to each other frame's point $\mathbf{x}$ and $\mathbf{x}'$ can be retrieved directly from $\mathbf{F}$ according to Eq. (2.14). Thus, for each point of a frame, the correspondence on the other one is searched for in the neighborhood of the direction of the epipole. Further details on two-view geometry can be found in Hartley and Zisserman (2003).

$$\mathbf{l}' = \mathbf{F}\mathbf{x} \qquad\qquad \mathbf{l} = \mathbf{F}^T \mathbf{x}' \qquad\qquad (2.14)$$

In order to solve find the depth of the point $\mathbf{X}$, the problem can be expressed as in Eq. (2.15), using Eq. (2.6), with $\mathbf{x}$ and $\mathbf{x}'$ the two projection of the 3D point $\mathbf{X}$. The camera are defined by their calibration matrix $\mathbf{K}$ and $\mathbf{K}'$. The relative pose between the two camera are defined by $\mathbf{R}$ and $\mathbf{t}$.

$$\mathbf{x} = \frac{1}{Z}\mathbf{K}[\mathbf{I}|\mathbf{0}]\mathbf{X} \qquad\qquad \mathbf{x}' = \frac{1}{Z'}\mathbf{K}'[\mathbf{R}|\mathbf{t}]\mathbf{X} \qquad\qquad (2.15)$$

Successively, the equality between the equivalent Eq. (2.16) is enforced by minimizing the loss function over Z and Z', according to Eq. (2.17). The system overconstrained by three equation and only two parameters. Finally, Z^* corresponds to the depth of $\mathbf{X}$ within the SRS of camera C, while Z'^* corresponds to the depth of $\mathbf{X}$ within the SRS of camera C'.

$$\mathbf{X} = Z\mathbf{K}^{-1}\mathbf{x} \qquad\qquad \mathbf{X} = \mathbf{R}^{-1}Z'\mathbf{K}'^{-1}\mathbf{x}' - \mathbf{R}^{-1}\mathbf{t} \qquad\qquad (2.16)$$

$$(Z^*, Z'^*) = \operatorname{argmin}_{Z, Z'} \|(Z\mathbf{K}^{-1}\mathbf{x}) - (\mathbf{R}^{-1}Z'\mathbf{K}'^{-1}\mathbf{x}' - \mathbf{R}^{-1}\mathbf{t})\|_2 \qquad\qquad (2.17)$$

The stereoscopy make use of the epipolar geometry to triangulate 3D points observed from two point of view. We now present the dense stereoscopy, the active stereoscopy and the structured-light theory used in 3D cameras.

2.2.3   Dense Stereoscopy

Dense Stereoscopy finds homologous keypoints to triangulate by detecting features and matching them coherently.

Although very simple in theory, a crucial problem consists in identifying which pixels correspond to the same 3D point across frames. This matching task has been widely addressed and the literature is wide on the problem. The main algorithm are split between *Feature-based matching* and *Correlation-based matching*.

Feature-based matching consists in detecting interesting points that can be univocally identified across frames. Such points are detected by looking for significant geometrical elements such as edges, curves and corners, depending on the geometrical properties of the scene. The points are matched based on a comparison between descriptive features, finding for similarity between the two sets of keypoints. Such approach has the main advantages of not being computational demanding and rather robust to intensity variations. On the other hand, it leads to a sparse disparity map, especially in environment providing a small amount of features.

Correlation-based matching is based on point-to-point correlation techniques between frames to identify corresponding pixels. A fixed-size window is defined around any point $\mathbf{P}$ of an image. The second image is correlated with this patch, in order to find a corresponding point $\mathbf{P}'$. A large set of correlation algorithms have been implemented in literature, adapting for different correlation function. Compared to the feature-based one, such approach is more computational demanding, sensitive to intensity variations (the matching frames must have the same intensity levels) and requires textured surfaces. On the other hand, it provides denser disparity maps, important aspect for shape reconstruction.

To improve and speed up the matching process, several constraints can be enforced. The most effective one is about the a-priori knowledge of the system's geometry: by knowing the geometrical relationships between frames, a point's correspondence in the other frames can be constrained along a proper direction. Epipolar geometry defines a point-to-line correspondence between projections on frames: given a two-cameras stereo system, the correspondences of a frame's points in the other one, stay along straight lines, reducing the starting 2D finding to a 1D problem. Applied to correlation-based approaches, it allows to bound the correlation process over lines, instead of processing all frame's points. In feature-based methods, it enforces matching only between detected frames' features, which satisfy such constraint.

2.2.4   Active Stereoscopy

Feature-based algorithms need scenes rich of geometrical features, while correlation-based ones need highly textured scenes. Whether dealing with features

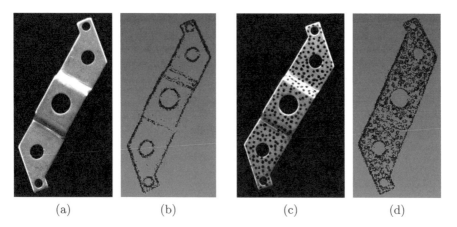

(a)   (b)   (c)   (d)

Fig. 2.12 Comparison between a textured part's (**c**) point cloud (**d**), and a non-textured part's (**a**) point cloud (**b**)

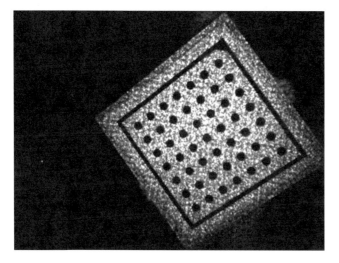

Fig. 2.13 Example of unstructured projected IR pattern (Intel Euclid)

matching or correlation matching algorithms, the density of a range map is function of the quantity of features and amount of texture available in the scene. Figure 2.12 shows two sample point clouds obtained capturing a non textured object (left side) and a textured one (right side), with a stereo system. Here, the textures has been painted only manually over the part. The low-texture object on the left have less interesting points to triangulate with compared to the high-textured on the right, resulting in a less dense point cloud.

In order to cope with scenes that lack features, a texture can be forced by artificially projecting a particular pattern over the scene. Figure 2.13 shows the infrared pattern projected by the Intel Euclid 3D camera. Active Stereoscopy technology projects a random light pattern that helps to triangulate on additional features. The pattern geometry has to be chosen carefully to avoid ambiguities in homologous points recognition. Too similar, too small or too closer geometries should be avoided. Theoretically, the most appropriate strategy would be a white noise texturing, where each pixel intensity level is independent of the surrounding ones. This would allow for denser patterns which lead to a better reconstruction. In practice, real applications exploit random dot patterns in the infrared domain to keep insensitive with respect to the ambient light.

2.2.5   Structured-Light

Structured-Light-based cameras sensors use a single camera with a structured pattern projected in the scene. Instead of triangulating with two cameras, a camera is substituted by a laser projector. It projects a codified pattern that embed enough structure to provide unique correspondences to triangulate with the camera. The direction of the structured pattern is known a priori by the camera, which is able to triangulate based on the pattern.

The most simple structured-light system is the laser triangulation shown in Fig. 2.14. A laser beam is projected on a scene or an object. The camera localizes the dot on the scene and recovers its position and depth following Eqs. (2.18) and (2.19). To improve the quality of the recognition, powerful IR laser are usually used.

$$z = \frac{b}{tan(\alpha) + tan(\beta)} \tag{2.18}$$

$$x = z \cdot tan(\alpha) \tag{2.19}$$

Such triangulation principle can be extend in a laser blade setting, where instead of a single dot, a laser plane intersects the shape to reconstruct, as shown in Fig. 2.15. A camera recognize the laser blade in the image and perform a triangulation for each point of the line. The sole equation of the laser plane within the camera SRS is enough to acquire the profile of any object in the scene. Note that such method can reach sub-millimeter accuracy.

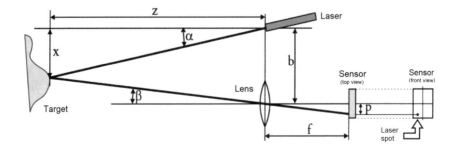

Fig. 2.14 Triangulation with a single laser spot

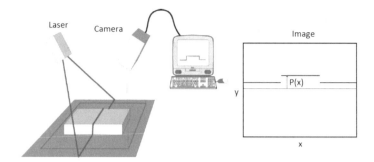

Fig. 2.15 Triangulation with a laser blade

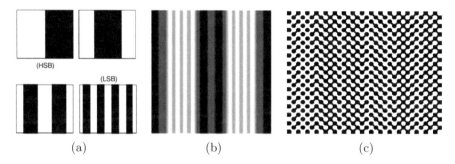

Fig. 2.16 *From left to right*: time-multiplexing strategy, direct coding and spatial neighborhood

Laser spot and laser blade triangulation respectively reconstruct the shape along 0 and 1 dimension. A structured-light laser system projects a pattern of codified light in 2 dimensions to estimate a dense range map. Such pattern is either time-coded, color-coded or spatially-coded. As such, according to Salvi et al. (2004), structured-light systems exploit either a *time-multiplexing strategy*, a *direct coding* or the *spatial neighborhood*.

Time-Multiplexing is the most common strategy when it comes to early structured-light. The projected pattern split the scene in several area of interest, creating edges to triangulate with. Figure 2.16a shows an example of time multiplexed patterns based on the Gray code. Due to the temporal aspect of the acquisition, *Time-Multiplexing* strategies do not allow for dynamic shape reconstruction. Nevertheless, they can provide dense shape acquisition with similar accuracy than with using a laser blade.

Direct coding makes use of grey scale or color coded pattern projected on the scene. The camera triangulates on particular textures the pattern projects on the scene, the which shape is known a priori. Figure 2.16b shows an example of gray-coded pattern proposed by Horn and Kiryati (1999). Such method permits a direct depth map measurement with a single frame, but is very sensitive to the surrounding light and may works only on dark scene, especially in case of colored pattern.

Spatial Neighborhood uses a spatially-structured pattern that creates uniqueness in the neighborhood of each projected pixel. Figure 2.16c shows the structured pattern proposed by Vuylsteke and Oosterlinck (1990) composed only of binary values. The Spatial Neighborhood approach allow for single frame reconstruction provided that the pattern is visible by the camera. Note that most of current commercial 3D cameras based on structured-light technology use a spatial neighborhood method in the IR bandwidth.

2.3   Depth by Time-of-Flight

Instead of estimating depth by observing the target from different prospective, the TOF principle can directly estimate the device-target distance.

2.3.1   Time-of-Flight Signal

Time-of-Flight (TOF) techniques have been employed for more than a century for ranging purposes. SONAR and RADAR are two techniques that exploit the sound and radio signals TOF principles, particularly in aerospace and aeronautic applications. More recently, with the improvement and the maturity of electronic devices, it has been possible to employ light signals for TOF systems. Applications using such system are numerous, especially in industrial and consumer fields.

The core of an optical Time-of-Flight system consists of a light transmitter and a receiver. The transmitter sends out a modulated signal that bounces off objects in the scene and senses back the reflected signal that returns to the receiver. The round-trip time from the transmitter to the receiver is an indicator of the distance of the object that the signal bounced back from. If the signal is periodic, the phase shift between the transmitted and the received signal can be used as an indicator of the round-trip time.

One of the simplest TOF systems is a single-pixel TOF system, also referred as *ranger*. A ranger provides the distance information for a single spot. Typically, IR or Near Infra-Red (NIR) light signals are used to reduce natural light interferences from the environment. Also, it is invisible to human eyes. Figure 2.17 illustrates

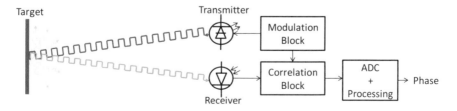

Fig. 2.17   Time-of-Flight emission, reflection and reception principle

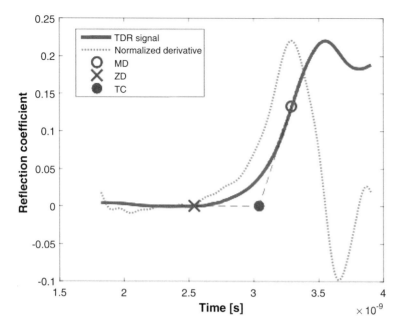

Fig. 2.18 POIs defined by Giaquinto et al. (2015)

the back-and-forth transmission of the light source signal through the environment to the target. The distance is obtained according to Eq. (2.20) from the time delay Δt and the light speed c. Note that the light travels 0.3 m/ns, which means that the estimation of the delay has to be very accurate.

In order to provide a depth map of an entire scene with a ranger, some sort of scanning must be performed. *Laser Scanners* are such systems, they typically orientate a TOF laser beam around two angles in order to reconstruct complete 3D environments. Such system measures up to a million of points in space per second.

$$d = \frac{c * \Delta t}{2} \qquad (2.20)$$

Single-Photon Avalanche Diodes (SPADs) or Geiger-mode Avalanche Photo Diodes (GAPDs) are commonly used to sense the received light signal very precisely. Those sensors have the capacity of capturing individuals photons with a very high resolution of a few tens of picoseconds (see Charbon et al. 2013), corresponding to a few millimeters of light traveling distance. Giaquinto et al. (2016) focus on finding the best Point of Interest (POI) in the step response, in order to determine the distance from the elapsed time between its emission and its reception according to Eq. (2.20). Those POI are commonly used in Time Domain Reflectometry applications to characterize and locate faults in metallic cables. Giaquinto et al. (2015) compared different POI such as the Maximum Derivative (MD), the Zero Derivative (ZD) and the Tangent Crossing (TC) criteria as shown in Fig. 2.18. They

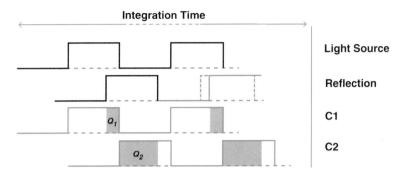

Fig. 2.19 Pulsed-modulation method from Li (2014)

point out that Tangent Crossing provides the best performances in term of systematic errors and repeatability.

A naive solution for time measurements would consist in a fast counter between POI of emitted and received signals, however signal processing provides better time estimation. Integrated systems measures distances exploiting the TOF principle use either *pulsed-modulation* or *Continuous-Wave (CW) modulation* (see Hansard et al. 2012).

The *pulse-modulation* method is straight-forward. It requires very short light pulses with fast rise- and fall-times, as well as high optical power like lasers or laser diodes. A solution is presented by Li (2014) and considers two out-of-phase windows C_1 and C_2, as shown in Fig. 2.19. It estimates the delay as the ratio of photons Q_2 that strikes back C_2 respect to the total energy $Q_1 + Q_2$ that strikes back both C_1 and C_2, as defined in Eq. (2.21). Note that it is limited to a 180° phase estimation and takes into account multiple periods over an integration time t_{int}. A well know measurement principle stands that using multiple measurement of Q_1 and Q_2 over the multiple periods improves the precision of such measurement. For that reason, *pulsed-modulation* measurement is more efficient than estimating the delay between two POI.

$$d = \frac{c * t_{int}}{2} \cdot \frac{Q_2}{Q_1 + Q_2} \qquad (2.21)$$

The *Continuous-Wave (CW) modulation* method uses a cross-correlation operation between the emitted and the received signal to estimate the phase between the two signals. This operation also takes into account multiple samples hence provides improved precision (see Hansard et al. 2012). The CW method modulates the emitted signal in a range of frequency of 10–100 MHz (see Dorrington et al. 2011). Ideally, square-wave signals are preferred, but different shapes of signals exist. Also, due to high-frequency limitation, transition phases during rise- and fall-times are significant. Cross-correlation between the emitted and the received signals permits a robust and more precise phase shift ϕ estimation, which returns a delay

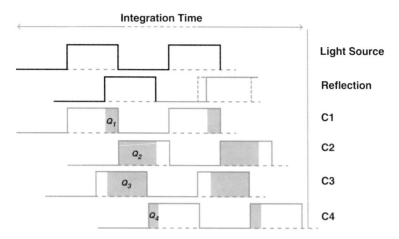

Fig. 2.20 Four phase-stepped samples according to Wyant (1982)

Δt hence a distance d quantity knowing the modulation frequency f according to Eq. (2.22).

$$d = \frac{c * \Delta t}{2} \quad \text{with} \quad \Delta t = \frac{\phi}{2\pi f} \tag{2.22}$$

Creath (1988) compared multiple phase-measurement methods, from which the *four-bucket* technique is the most widely used. Wyant (1982) originates this technique that takes into consideration four samples of the emitted signal, phase-stepped by 90°, as shown on Fig. 2.20. Electrical charges from the reflected signal accumulates during these four samples and the quantity of photons are probed in Q_1, Q_2, Q_3 and Q_4. Creath (1988) presented the *four-bucket* method estimates the phase shift according to Eq. (2.23).

$$d = \frac{c}{4\pi f}\phi \quad \text{with} \quad \phi = \text{atan}\left(\frac{Q_3 - Q_4}{Q_1 - Q_2}\right) \tag{2.23}$$

Looking closer at the CW phase ϕ, we can notice that the differences $Q_3 - Q_4$ and $Q_1 - Q_2$ normalize any constant offset in the returned signal. Offset occurs when environmental light interferes with the transmitted signal. Also, the ratio between $(Q_3 - Q_4)$ and $(Q_1 - Q_2)$ provides a normalization for the amplitude. Actually, the quantity of energy received is reduced respect to the emitted one, due to dispersion, which yields to an amplitude reduction. Being independent of both signal offset and attenuation is necessary for a robust phase estimation. The *four-bucket* method also provides amplitude (A) and offset (B) estimations of the returned signal following the Eqs. (2.24).

$$A = \frac{\sqrt{(Q_3 - Q_4)^2 + (Q_1 - Q_2)^2}}{2}$$

$$B = \frac{Q_1 + Q_2 + Q_3 + Q_4}{4}$$

(2.24)

According to Li (2014), the amplitude A and the offset B of the reflected signal influence the depth measurement accuracy σ. The measurement variance can be approximated by Eq. (2.25), where the modulation contrast c_d describes how well the TOF sensor separates and collects the photoelectrons. It is worth noting that high amplitude, high modulation frequency (up to a physical limit) and high modulation contrast actually improve the accuracy. Also, Li (2014) shows that a large offset can lead to saturation and inaccuracy.

$$\sigma = \frac{c}{4\sqrt{2}\pi f} \cdot \frac{\sqrt{A + B}}{c_d A}$$

(2.25)

The range of measurement is limited by positive value up to an *ambiguity distance* corresponding to a 2π phase shift, after which a wraparound occurs. This ambiguity distance is defined in Eq. (2.26). With a single frequency technique, the ambiguity distance can only be extended reducing the modulation frequency and, as a consequence, accuracy reduces (see Eq. (2.25)). For a 30 MHz modulation frequency, the unambiguity ranges from 0 to 5 m.

$$d_{amb} = \frac{c}{2f}$$

(2.26)

Advanced TOF systems deploy multi-frequency technologies, combining more modulation frequencies. The combination of multiple frequencies in the signal increases its period, being the least common multiple of the two component periods. A dual-frequency concept is illustrated in Fig. 2.21. The *beat frequency* is defined as the frequency when the two modulations agree, corresponding to the new ambiguity distance, usually higher than single frequency technology. Multi-frequency systems can reach kilometers range of performance.

Precision in the depth measurement is provided by the phase estimation uncertainty. Assuming it follows a normal distribution, since the distance estimation d is linear with the phase ϕ in Eq. (2.23), the distance error distribution will also follow a normal distribution. Figure 2.22a illustrates the linear propagation of the uncertainty.

Low-frequency f_0 provides larger range of measurement (Eq. (2.26)), but yields a large depth uncertainty (see Fig. 2.22a). Figure 2.22b shows a higher frequency f_1 operation, in which the phase wraparound takes place at shorter distances, causing more ambiguity in the distance measurement. However, depth uncertainty is smaller for a given phase uncertainty. Figure 2.22c shows another higher frequency f_2 operation. Using two frequencies, as shown in Fig. 2.22b, c, it is possible to disambiguate the distance measurement by picking a consistent depth value across

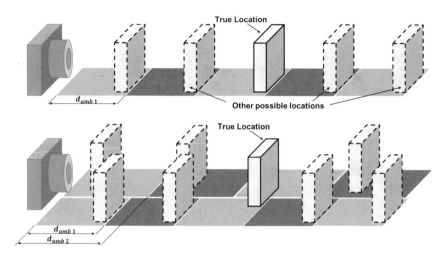

Fig. 2.21 *Top*: Single frequency aliasing phenomenon. *Bottom*: Dual frequency extension of the range of measurement

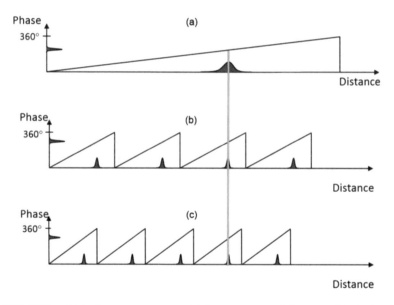

Fig. 2.22 TOF multiple frequency operation. (**a**) Low-frequency f_0 has no ambiguity but large depth uncertainty. (**b**) High-frequency f_1 has more ambiguity but small depth uncertainty. (**c**) Another high-frequency f_2 with small depth uncertainty

the two frequencies. The depth values for f_1 and f_2 are then averaged together to produce an even smaller depth uncertainty that f_1 or f_2 alone. This uncertainty is approximately the uncertainty of the average frequency of f_1 and f_2 applied to the total integration time required.

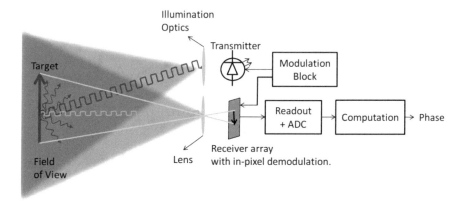

Fig. 2.23 Time-of-Flight camera model

2.3.2 Time-of-Flight Cameras

TOF cameras are depth sensors that exploit the TOF principle for every pixel of the matrix sensor. TOF cameras are 3D sensors that estimates depth with a direct range measurement.

In a TOF camera, the transmitter consists of an illumination block that illuminates the region of interest with modulated lights, while the sensor consists of a matrix of pixels that collects light from the same region of interest, as shown in Fig. 2.23. TOF sensors usually have in-pixel demodulation, that is, each pixel develops a charge that represents the correlation between transmitted and received light. A lens is required for image formation while maintaining a reasonable light collection area because individual pixels must collect light from distinct parts of the scene.

2.4 From Depth Map to Point Cloud

With the maturity of 2D computer vision, range images are convenient to use. Softwares and libraries like Matlab, HALCON or OpenCV already include several tools for image elaboration. Nevertheless, point cloud and meshes are more suitable for general 3D shape representation. 3D cameras are often coupled with RGB sensors, in order to add color information to the depth. Such device are defined as *RGB-D cameras*, providing 4-channels images. In the following, we show how to transform a depth map in a 3D point cloud and how to registrate the color.

3D data are more complex to manage, since a third dimension is added to the data. Images are structured matrices while point clouds are 3D scattered data. Pixel sampling is also a convenient operation since the data are picked in a limited resolution. In 3D, the number of point are infinite and data are usually stored in

Fig. 2.24 Depth
measurement

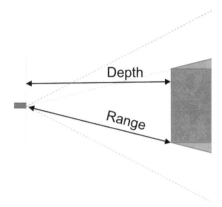

three floats representing the three components along the three axes. Nevertheless,
3D information are richer than 2D and every single point on space can be stored,
even if occluded from a given point of view. For this reason, and especially for 3D
reconstruction, 3D data storing is essential.

3D cameras provide depth frames, regardless of the technology employed,
that usually return the distance measurements between the target and the image
plane along the optical axis. Note that to obtain this measurement, a minimum
of information about the intrinsic parameters are necessary such as focal length,
optical center and distortion, but RGB-D cameras usually perform it independently
and provide range maps along the optical axis. Figure 2.24 presents such depth
measurement.

A point cloud is a data structure used to represent a collection of multi-
dimensional points. Commonly, a point cloud is a three-dimensional set that
encloses the spatial coordinates of an object sampled surface. However, geometrical
or visual attributes can be added to each point. Using a range map, depth measurements are reprojected in 3D. A 3D point $\mathbf{M}$ with coordinates (X, Y, Z) is obtained
according to Eq. (2.27) from the depth information $\mathbf{D}_{x,y}$, (x, y) being the rectified
pixel position on the sensor.

$$
\begin{aligned}
X &= \mathbf{D}_{x,y} * (c_x - x)/\mathbf{f}_x \\
Y &= \mathbf{D}_{x,y} * (c_y - y)/\mathbf{f}_y \\
Z &= \mathbf{D}_{x,y}
\end{aligned}
\tag{2.27}
$$

Those equations are not linear due to the non linearity of x and y estimation
introduced by the non linear camera model. In order to improve 2D to 3D conversion
speed, lookup tables are usually used. Two lookup tables store coefficients that,
multiplied by the depth of a given pixel, return the X and the Y values of the point
$\mathbf{M}$ in space. As a result, point clouds are produced from a depth map as shown in

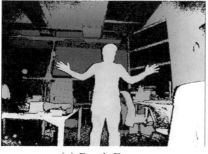

(a) Depth Frame   (b) Point Cloud

Fig. 2.25 Point cloud (**b**) obtained from the depth map (**a**)

(a) Colored depth frame   (b) Colored point cloud

Fig. 2.26 Color Point cloud (**b**) obtained from the colored depth map (**a**)

Fig. 2.25. The transformation, being a simple multiplication, can be perform very efficiently on a Graphics Processing Unit (GPU).

Since 3D sensors are often coupled with an RGB camera, we investigate how the color registration on depth operates. Registering two cameras means knowing the relative position and orientation of a SRS respect to another. In essence, the idea behind color integration consists in re-projecting every 3D points on the RGB image, in order to adobe its color. When reprojected in 3D, the generated point cloud contains six information fields, three of them are space coordinates while the remaining three are color coordinates. Note that not all the 3D points reconstructed on the scene are visible from the RGB camera, some points may lack color information due to occlusion. Figure 2.26 shows the result of a colorization of the previous depth map and previous point cloud.

Chapter 3
State-of-the-Art Devices Comparison

In this section, we present a non-exhaustive list of the most important Three-dimension (3D) camera sensors, devices and solutions available for the mass market. An overview of the main characteristics is provided in Table 3.1.

3.1 PMD Technologies

PMD Technologies is a German developer of Time-of-Flight (TOF) components and a provider of engineering supports in the field of digital 3D imaging. The company is named after the Photonic Mixer Device (PMD) technology used in its products to detect 3D data in real time. They have been famous in the early 2000s for the first TOF devices available for research purposes, such as the PMD CamCube^{TM} (Fig. 3.1). More recently, they presented the PMD CamBoard^{TM} pico flexx (Fig. 3.2) which aims to the consumer market with a smaller scale factor.

3.2 MESA Imaging

MESA Imaging is a company founded in July 2006 as a spin out from the Swiss Center for Electronics and Microtechnology (CSEM) to commercialize its TOF camera technologies. They propose two generations of 3D TOF cameras, the SwissRanger 4000^{TM} (Fig. 3.3) and the SwissRanger 4500^{TM} (Fig. 3.4). Both device were widely used for research purposes. In 2014, *MESA Imaging* was bought by *Heptagon*.

© The Author(s), under exclusive licence to Springer International Publishing AG, part of Springer Nature 2018
S. Giancola et al., *A Survey on 3D Cameras: Metrological Comparison of Time-of-Flight, Structured-Light and Active Stereoscopy Technologies*, SpringerBriefs in Computer Science, https://doi.org/10.1007/978-3-319-91761-0_3

Table 3.1 Comparison of the main 3D camera commercially available

Device	Technology	Range (m)	Resolution	Frame rate (fps)	Field of view
PMD CamCube 2.0[TM]	Time-of-Flight	0–13	200 × 200	80	40° × 40°
PMD CamBoard[TM]	Time-of-Flight	0.1–4.0	224 × 171	45	62° × 45°
MESA SR 4000[TM]	Time-of-Flight	0.8–8.0	176 × 144	30	69° × 56°
MESA SR 4500[TM]	Time-of-Flight	0.8–9.0	176 × 144	30	69° × 55°
ASUS Xtion[TM]	Structured-light	0.8–4.0	640 × 480	30	57° × 43°
Occipital[TM]	Structured-light	0.8–4.0	640 × 480	30	57° × 43°
Sense 3D scanner[TM]	Structured-light	0.8–4.0	640 × 480	30	57° × 43°
Kinect V1[TM]	Structured-light	0.8–4.0	640 × 480	30	57° × 43°
Kinect V2[TM]	Time-of-Flight	0.5–4.5	512 × 424	30	70° × 60°
Creative Senz 3D[TM]	Time-of-Flight	0.15–1.0	320 × 240	60	74° × 58°
SoftKinetic DS325[TM]	Time-of-Flight	0.15–1.0	320 × 240	60	74° × 58°
Google Tango[TM] Phone	Time-of-Flight	–	–	–	–
Google Tango[TM] Tablet	Structured-light	0.5–4.0	160 × 120	10	–
Orbbec Astra S[TM]	Structured-light	0.4–2.0	640 × 480	30	60° × 49.5°
Intel SR300[TM]	Structured-light	0.2–1.5	640 × 480	90	71.5° × 55°
Intel R200[TM]	Active stereoscopy	0.5–6.0	640 × 480	90	59° × 46°
Intel Euclid[TM]	Active stereoscopy	0.5–6.0	640 × 480	90	59° × 46°
Intel D415[TM]	Active stereoscopy	0.16–10	1280 × 720	90	63.4° × 40.4°
Intel D435[TM]	Active stereoscopy	0.2–4.5	1280 × 720	90	85.2° × 58°
StereoLabs ZED[TM]	Passive stereoscopy	0.5–20	4416 × 1242	100	110°($diag.$)

Device	PMD CamCube 2.0
Technology	Time-of-Flight
Range	$0 - 13.0$ m
Resolution	200×200 pix
Frame Rate	up to 80 fps
Field of View	$40° \times 40°$

Fig. 3.1 Presentation and characteristics of the PMD CamCube 2.0[TM]

Device	PMD CamBoard
Technology	Time-of-Flight
Range	$0.1 - 4.0$ m
Resolution	224×171 pix
Frame Rate	up to 45 fps
Field of View	–

Fig. 3.2 Presentation and characteristics of the PMD CamBoard[TM]

Device	MESA SwissRanger 4000
Technology	Time-of-Flight
Range	0.8 − 8.0 m
Resolution	176 × 144 pix
Frame Rate	30fps
Field of View	69° × 56°

Fig. 3.3 Presentation and characteristics of the MESA SwissRanger 4000™

Device	MESA SwissRanger 4500
Technology	Time-of-Flight
Range	0.8 − 9.0 m
Resolution	176 × 144 pix
Frame Rate	30fps
Field of View	69° × 55°

Fig. 3.4 Presentation and characteristics of the MESA SwissRanger 4500™

Device	ASUS Xtion
Technology	Structured-Light
Range	0.8 − 4.0 m
Resolution	640 × 480 pix
Frame Rate	30 fps
Field of View	57° × 43°

Fig. 3.5 Presentation and characteristics of the ASUS Xtion™

3.3 PrimeSense

PrimeSense is an Israeli company that produce structured-light chips for 3D cameras. They manufacture a sensor able to acquire at a 640 × 480 pixel resolution, even though the real spatial resolution is 320 × 240. Such sensor is used, among other, in the *ASUS Xtion*™ (Fig. 3.5), the *Occipital*™ (Fig. 3.6) and the *Sense 3D scanner*™ (Fig. 3.7). *PrimeSense* contributed to the democratization of 3D camera by providing such low-cost structured-light chips. In 2013, *PrimeSense* was purchased by *Apple*.

3.4 Microsoft Kinect

Microsoft is the company that help the most in distributing 3D camera to a large public. In November 2010, they brought millions of 3D camera in gamers living rooms, by releasing the first version of the Kinect (Fig. 3.8). The Kinect is a 3D

Device	Occipital
Technology	Structured-Light
Range	0.8 − 4.0 m
Resolution	640 × 480 pix
Frame Rate	30 fps
Field of View	57° × 43°

Fig. 3.6 Presentation and characteristics of the Occipital™

Device	Sense 3D scanner
Technology	Structured-Light
Range	0.8 − 4.0 m
Resolution	640 × 480 pix
Frame Rate	30 fps
Field of View	57° × 43°

Fig. 3.7 Presentation and characteristics of the Sense 3D scanner™

Device	Kinect V1
Technology	Structured-Light
Range	0.8 − 4.0 m
Resolution	640 × 480 pix
Frame Rate	30 fps
Field of View	57° × 43°

Fig. 3.8 Presentation and characteristics of the Kinect V1™

Device	Kinect V2
Technology	Time-of-Flight
Range	0.5 − 4.5 m
Resolution	512 × 424 pix
Frame Rate	30 fps
Field of View	70° × 60°

Fig. 3.9 Presentation and characteristics of the Kinect V2™

camera used to interact with the Xbox360 with the player body. It is build by the company *PrimeSense* and integrates state-of-the-art algorithm to track up to six human body in a scene (Shotton et al. 2011).

A couple of years later in 2012, Microsoft released a specific version for Windows, with an SDK providing tools for human body and face tracking. Additionally, the Kinect uses a color camera registered with the structured light depth system, as well as a set of microphone for sound localization.

In fall 2013, *PrimeSense* was bought by *Apple* and *Microsoft* presented the second version of the *Kinect* for its new XBoxOne gaming console (Fig. 3.9). The sensor is based on a proprietary design, presented by Bamji et al. (2015). The TOF sensor is a $0.13\,\mu m$ CMOS system-on-chip with a resolution of 424×512 pixels, the highest for a TOF cameras. It provides measurement up to 4.5 m with the official SDK, but open-source library such as libfreenect2 enlarges the range up to 12.5 m. The Kinect V2^{TM} also includes a 1080p RGB sensor, registered with the TOF camera.

3.5 Texas Instrument OPT8140

The Creative Senz3D^{TM} and the SoftKinectic DS325^{TM} may are the first consumer-oriented TOF cameras, designed primarily as input devices for computer entertainment. They are built by different manufacturer, but the TOF sensor is identical, an OPT8140 provided by *Texas Instrument*. The CMOS sensor has a 320×240 pixels resolution and a modulation frequency ranging from 0 to 80 MHz. Depth measurements range from 0.15 to 1 m and can reach up to 60 fps. Both of the devices are flanked with a 720×1280 pixel RGB camera, calibrated with the depth camera.

Creative Technology is a company based in Singapore, that released in June 2013 the Creative Senz3D^{TM} TOF camera in collaboration with *Intel* and its *Real Sense* philosophy. The Creative Senz3D^{TM} has been designed to enhance personal computer interaction by introducing gesture control. Typically, this sensor enables hand and finger tracking as well as face detection and recognition (Fig. 3.10).

SoftKinetic is a Belgian company that develops TOF solutions for consumer electronics and industrial applications. In June 2012, the firm presents their SoftKinectic DS325^{TM} TOF device and provides their own gesture recognition software platform, named *iisu*, that allows natural user interface on multiple Operative Systems. *SoftKinetic* was bought by *Sony Corporation* in October 2015 (Fig. 3.11).

Device	Creative Senz 3D
Technology	Time-of-Flight
Range	$0.15 - 1$ m
Resolution	320×240 pix
Frame Rate	60 fps
Field of View	$74° \times 58°$

Fig. 3.10 Presentation and characteristics of the Creative Senz 3D^{TM}

Device	SoftKinetic DS325
Technology	Time-of-Flight
Range	$0.15 - 1$ m
Resolution	320×240 pix
Frame Rate	60 fps
Field of View	$74° \times 58°$

Fig. 3.11 Presentation and characteristics of the SoftKinetic DS325^{TM}

Device	Google Tango Tablet
Technology	Structured-Light
Range	$0.5 - 4$ m
Resolution	160×120 pix
Frame Rate	10 fps
Field of View	—

Fig. 3.12 Presentation and characteristics of the Google Tango^{TM} Tablet

Device	Google Tango Phone
Technology	Time-of-Flight
Range	—
Resolution	—
Frame Rate	—
Field of View	—

Fig. 3.13 Presentation and characteristics of the Google Tango^{TM} Phone

3.6 Google Tango^{TM}

In 2014, *Google* launched the project *Tango*, a mobile platform that brings augmented reality features to mobile devices like smartphone and tablets. The depth sensors for the Google Tango^{TM} Tablet (Fig. 3.12) are manufactured by *OmniVision*, a Taiwanese company. The tablet embeds a NVIDIA Tegra K1 processor with 192 CUDA cores as well as 4 GB of RAM and 128 GB of internal storage, which makes it the most portable and compact solution available. This opens up many augmented/virtual reality, motion tracking, depth perception and area learning applications. Only a few official information were released by Google.

Google also presented a smartphone version of its Tango device, based on a TOF sensor manufactured by *PMD Technologies* (Fig. 3.13). In June 2016, *Lenovo* presented the *Phab 2 Pro^{TM}*, the first consumer-based phone with a 3D camera embedded. In August 2017, *Asus* presented the *Zenfone AR^{TM}*, based on the same technology.

Device	Orbbec Astra S
Technology	Structured-Light
Range	$0.4 - 2.0$ m
Resolution	640×480 pix
Frame Rate	30 fps
Field of View	$60° \times 49.5°$

Fig. 3.14 Presentation and characteristics of the Orbbec Astra S^{TM}

As for March 2018, the Google Tango^{TM} Project have been deprecated, in favor of the development of the Google Augmented Reality tool ARCore.

3.7   Orbbec

Orbbec is a company founded in China, manufacturing 3D cameras based on the structured light technology. The propose the Astra S^{TM} camera (Fig. 3.14), composed of an IR camera, a coded pattern projector and a RGB camera. The device includes also two microphones and a proximity sensor. Orbbec released an SDK allowing for human skeletal recognition and tracking, similarly to what Microsoft did with the Kinect.

Orbbec also presented the Persee^{TM}, based on the Orbbec Astra^{TM}. They enclose a quad-core 1.8 GHz ARM processor (A17 family), a 600 MHz Mali GPU, Ethernet and Wi-Fi network connections directly within their 3D camera. They propose a convenient solution for embedded applications.

3.8   Intel RealSense^{TM}

Intel recently became an important actor in 3D camera, succeeding the fame of the Kinect generation. Intel RealSense^{TM} provides an open platform to exploit any 3D perceptual devices they produce. The LibRealSense^{TM} cross-platform APIs provide several tools for to manage sensor's streams, generate and process 3D point clouds, as well as advanced functions for hand tracking, fine face recognition and 3D scanning. Intel released various vision modules either based on active stereoscopy and structured-light technologies. In the following, we present the last generation of Intel devices.

In 2016, Intel released the *SR300* camera module (Fig. 3.15). The structured-light device is composed of an IR pattern projector, based on a system of resonating MEMS mirrors and lenses that diffuse the IR laser in a specific pattern. An imaging Application-Specific Integrated Circuit (ASIC) that performs the in-hardware depth computation and drops synchronized VGA-IR and a FullHD-RGB frames at up to 60 fps. The SR300 is designed for short range; preliminary test performed by

Device	Intel SR300
Technology	Structured-Light
Range	$0.2 - 1.5$ m)
Resolution	640×480 pix
Frame Rate	$30, 60, 90$ fps
Field of View	$71.5° \times 55°$

Fig. 3.15 Presentation and characteristics of the Intel SR300™

Device	Intel R200
Technology	Active Stereoscopy
Range	$0.5 - 6.0m$
Resolution	640×480 pix
Frame Rate	$30, 60, 90$ fps
Field of View	$59° \times 46°$

Fig. 3.16 Presentation and characteristics of the Intel R200™

Device	Intel ZR300/Euclid
Technology	Active Stereoscopy
Range	$0.5 - 6.0m$
Resolution	640×480 pix
Frame Rate	$30, 60, 90$ fps
Field of View	$59° \times 46°$

Fig. 3.17 Presentation and characteristics of the Intel ZR300/Euclid™

Carfagni et al. (2017) show good results between 0.2 m and 1.5 m, with optimal results within 0.7 m. Within such range, this device can be used effectively as a 3D scanner and is well suitable to gesture-based interfaces and 3D scanning of small objects.

The *R200™* module is a depth camera based on infrared active stereoscopy technology (Fig. 3.16). The depth is estimated in-hardware through an imaging ASIC that processes the infrared streams (together with the RGB one), performs frames correlation with a census cost function to identify homologous points and reconstructs the disparity. Since it is based on active stereoscopy, an infrared dot-pattern projector adds textures to the scene, to cope with low-texture environments. The main purposes of such a device are face recognition, gesture tracking and autonomous navigation. More details on this 3D camera is provided by Intel in Keselman et al. (2017).

Similar than the Orbbec Persee™ camera, Intel deployed the *Euclid* solution for the robotics and the intelligent devices fields (Fig. 3.17). It consists of a *ZR300* depth camera module, identical to the *R200™*, coupled with an embedded computer powered by an Atom X7-Z8700. From the hardware point of view, the ZR300 is a *R200™* depth camera module (Infra-Red (IR) stereo + RGB camera), coupled

Device	Intel D415
Technology	Active Stereoscopy
Range	$0.16 - 10$ m
Resolution	1280×720 pix
Frame Rate	$30, 60, 90$ fps
Field of View	$63.4° \times 40.4°$

Fig. 3.18 Presentation and characteristics of the Intel D415^{TM}

Device	Intel D435
Technology	Active Stereoscopy
Range	$0.2 - 4.5$ m
Resolution	1280×720 pix
Frame Rate	$30, 60, 90$ fps
Field of View	$85.2° \times 58°$

Fig. 3.19 Presentation and characteristics of the Intel D435^{TM}

with a tracking module including a fisheye camera and an IMU. The Euclid comes with ROS on-board and is able to interface with Arduino modules. The ZR300 module is provided by a whole set of sensors consisting of a VGA-IR stereo camera, a monochrome VGA fisheye camera (160 FOV), a rolling-shutter RGB (FullHD) camera and an IMU (3-axis accelerometer and 3-axis gyroscope). With such a sensing equipment, it enables a widespread set of perceiving applications ranging from autonomous navigation and robust tracking (exploiting sensors fusion approaches), to virtual/augmented reality, on a single device.

Intel's D400^{TM} serie is the evolution of the R200^{TM} serie. Under this generation, Intel released the *D415^{TM}* (Fig. 3.18) and the *D435^{TM}* (Fig. 3.19), the former featuring a rolling shutter and the latter a global shutter. Compared to R200^{TM} they are based on the same infrared active stereoscopy technology, with a better depth resolution of 1280×720 pixels. The depth estimation is still performed in-hardware by a specific ASIC, but the matching algorithms between infrared frames is drastically improved. Preliminary tests have shown that even indoor, without projector and poor textured objects, the cameras provide sufficient depth maps, whereas R200^{TM}-based ones show large holes in the depth map. Such algorithms are based on a correlation matching approach as reported by Intel in Keselman et al. (2017).

3.9   StereoLabs ZED^{TM}: Passive Stereo

Stereolabs is a French company proposing a passive stereoscopic system using only by the natural textures from the scene to infer depth. Their product ZED (Fig. 3.20) look promising, in particular in term of range, resolution and frame rate. In natural environment, with a large amount of texture, the device is capable of appreciate

Device	StereoLabs ZED
Technology	Passive Stereoscopy
Range	$0.5 - 20m$
Resolution	up to 4416×1242 pix
Frame Rate	up to 100 fps
Field of View	$110° (diagonal)$

Fig. 3.20 Presentation and characteristics of the StereoLabs ZED^{TM}

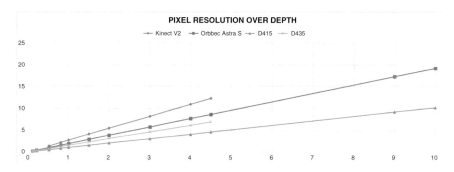

Fig. 3.21 Resolution and range synthesis for the main 3D cameras

depth. Nevertheless, in case of low texture, the passive stereoscopy technology is not able to reconstruct properly a scene and errors in the depth estimation may reach 1 m.

3.10 Discussion

Several 3D camera based on different technology have been developed during the last decade. The first 3D cameras were based on TOF principle (PMD, MESA), they were expensive 3D camera reserved exclusively for research or professional purposes. With the appearance of Structured-Light systems based on PrimeSense chips (ASUS Xtion^{TM}, Occipital^{TM}, Sense 3D^{TM}, Kinect V1^{TM}), the cost of 3D cameras drops drastically and such technology spread in to the consumer mass market. TOF device catched up later, in particular with the second iteration of the Kinect V2^{TM} that show state-of-the-art performance and a wide range of applications. Nevertheless, TOF system drawback relies on its high-energy consumption, powerful LEDs are required to spread the TOF signal around he scene. Recently, *Intel* is pushing forward the Active Stereoscopy technology, that provide decent depth estimation with low-energy consumption.

In the following we analyze the three main technologies that mark the last decade: TOF, Structured-Light and Active Stereoscopy. For the TOF technology, we analyse the Kinect V2^{TM} which present the best specifications (Chap. 4). For the Structured-Light technology, the most advanced and promising device is the

Orbbec Astra S^{TM} (Chap. 5). Regarding the Active Stereoscopy technology, the Intel D400 camera is the most recent device, providing promising resolution (Chap. 6). Figure 3.21 shows a comparison of the resolution for the three main devices we are investigating.

Chapter 4
Metrological Qualification of the Kinect V2^{TM} Time-of-Flight Camera

The Kinect V2^{TM} is a Time-of-Flight (TOF) camera device with state-of-the-art performances. Including the first version of the device, Microsoft sold tens of million of Kinects, proposing appealing low-cost Three-dimension (3D) cameras below 200€. The main specifications of the Microsoft Kinect V2^{TM} are summarized in Table 4.1. Bamji et al. (2015) released a full description of the 512×424 CMOS IR TOF sensor included in the Kinect V2^{TM}. The Kinect V2^{TM} also incorporates a full HD RGB camera, calibrated with the aforementioned depth sensor, and provide colored depth maps and point clouds at roughly 30 Hz. In this chapter, we investigate the performances of the Kinect V2^{TM} as a depth camera, focusing on uncertainty characterization according to the Guide to the Expression of Uncertainty in Measurement (GUM) (BIPM et al. 2008). First of all, the TOF signal transmitted by the Kinect V2^{TM} is evaluated. Then, stability is discussed as well as distribution normality. Range measurement uncertainty is studied at pixel and sensor scales. Last, qualitative results are provided in simple scenarios.

4.1 Time-of-Flight Modulated Signal

In this first section, we are verifying the modulated light signal components transmitted by the Kinect V2^{TM}. The Kinect V2^{TM} is composed of three LEDs that emit at a 827–850 nm Infra-Red (IR) wavelength. In order to measure the modulation frequencies of such signal to determine the ambiguity distance, we use an external photo-diode. The PDA10A-EC is a fixed gain detector provided by *ThorLabs*. This sensor has a 0.8 mm^2 active area, is sensible to wavelength from 200 to 1100 nm and provides intensity measurements up to a 150 MHz frequency range. The detector acquisition frequency range is enough to measure the Kinect V2^{TM} signal, since we are expecting the frequency components to be around 16 and 120 MHz. In order

© The Author(s), under exclusive licence to Springer International Publishing AG, part of Springer Nature 2018
S. Giancola et al., *A Survey on 3D Cameras: Metrological Comparison of Time-of-Flight, Structured-Light and Active Stereoscopy Technologies*, SpringerBriefs in Computer Science, https://doi.org/10.1007/978-3-319-91761-0_4

Table 4.1 Kinect V2^{TM}:
main characteristics

IR camera resolution	512 × 424	(pix)
RGB camera resolution	1080 × 1920	(pix)
Maximum frame rate	30	(Hz)
Field of View (FOV)	70(H) × 60(V)	(°)
Measurement range	500–4500	(mm)
Dimension	250 × 66 × 67	(mm)
Weight	966	(g)
Connection	USB 3.0 + power supply	
Operating system	Windows 8/10, Linux	
Software	Kinect V2 SDK, libfreenect2	

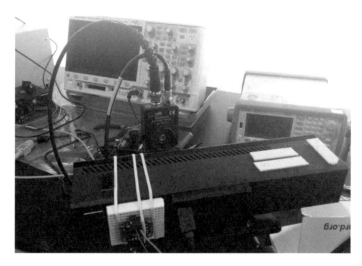

Fig. 4.1 Acquisition setup including the PDA10A-EC detector, the 4 GHz oscilloscope and the Kinect V2^{TM}

to acquire data in a convenient frame rate and without aliasing, we use a 4 GHz oscilloscope from *Keysight* technologies with data logger. The Kinect V2^{TM} has been placed pointing in direction of the detector while switched on. A complete setup composed of the detector, the oscilloscope and the Kinect V2^{TM} is presented in Fig. 4.1.

Two different tests are realized in order to verify the characteristics of Kinect V2^{TM} transmitted signal. First of all, the 30 Hz acquisition is verified. Then, the modulated signal is investigated. Figure 4.2 shows the acquired signal focusing on the 30 Hz frequency. It is a 100 ms acquisition at 20 kHz, in order to verify the Kinect V2^{TM} frame rate. As excepted, the acquisition frequency is equal to 30 Hz. The small spikes at 10 Hz is due to the frequency resolution given the 100 ms acquisition.

In the second test, we focus on the modulation frequencies. The integration time in Fig. 4.2 is actually split in three area respectively carrying the signal at 80 MHz, 16 MHz and 120 MHz. We have acquired samples at 4 GHz on these three area and

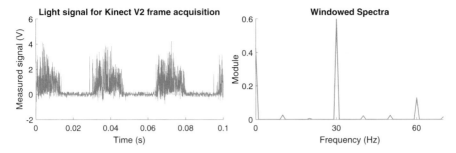

Fig. 4.2 30 Hz signal for grabbing along a 50% period integration time

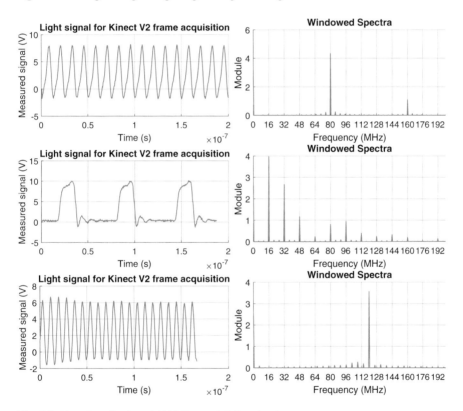

Fig. 4.3 80 MHz, 16 MHz and 120 MHz modulation signals

show the results in Fig. 4.3. First of all, it is worth noting that the signal is nor squared nor sinusoidal, due to the difficulties to alternatively switch on and off the LEDs and to perfectly control such phenomena at such a high frequency. Regarding the spectra, we show a first 80 MHz frequency modulation for 8 ms, followed by a frequency modulation of 16 MHz for 4 ms and a last frequency modulation of 120 MHz for 8 ms.

Rated voltage	12 V
Input current	0.04 A
Speed	5000 rpm
Max air volume	0.0026 m^3/s

Fig. 4.4 Detail and general specifications of the fan used as external cooling system

Note that in this test the Kinect V2^{TM} device is controlled with the libfreenect2 library, setting a maximum range measurement to its default value, 4.5 m. Bamji et al. (2015) stands that frequencies tuning is possible on their original 0.13 μm system-on-chip sensor, but there is no clue that neither the original Software Development Kit (SDK) nor the libfreenect2 library are able to set such frequencies variables.

4.2 Temperature and Stability

The second step in this metrological qualification consists in verifying that the depth measurement is stable to environmental conditions. Electronic sensors and signal conditioning circuits are sensitive to temperature, that often causes output drifts. Since it has been noted that the Kinect V2^{TM} gets warmer after some minutes of activity, we have to verify the stability of the device output during static measurements.

We have noted that a fan is located inside the Kinect V2^{TM} device. This fan is controlled by a thermostat and switched on and off automatically when the device reaches a threshold temperature. The primary test consists in investigating the stability of the central pixel measurement with and without a continuous air flow.

As a matter of fact, it is not possible to manually turn on or off the internal cooling system. In order to maintain a continuous air flow, an external fan is fixed over the original one (Fig. 4.4). Indeed, the continuous rotation of the external cooling system sets the internal temperature under the low thermostat threshold, as to prevent the activation of the controller and rotation of the internal fan.

A first test is carried out acquiring 20, 000 samples at 30 Hz, placing the sensor at about one meter from a white planar wall. In order to highlight the measure trend, a moving average is calculated on 500 distance samples returned by the central pixel of the sensor. For the entire duration of this test (10 min), the internal cooling system remained off, because the temperature of the sensor remained below the high level threshold. A second 20, 000 samples acquisition was then performed with the

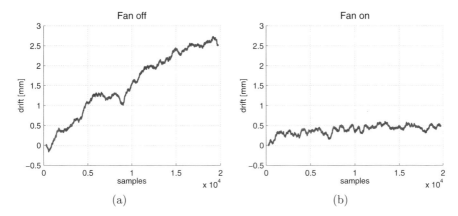

(a)   (b)

Fig. 4.5 Static measurements of a single pixel in time without (**a**) and with (**b**) cooling system. 20, 000 measurements were sampled at 30 Hz, during around 10 min

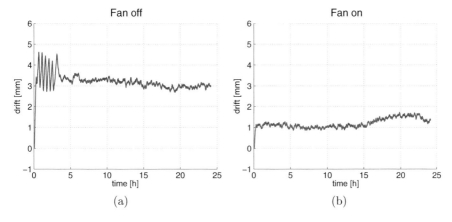

(a)   (b)

Fig. 4.6 Static measurements of a single pixel in time during 24 hours without (**a**) and with (**b**) cooling system

external fan switched on, using the same setup of the previous test. The central pixel distance data and the moving average of the two tests are plotted in Fig. 4.5.

While the figure on the left shows an increase of the mean value of more than 2 mm in 10 min, the test carried out with a continuous cooling system (right) was able to maintain a stable temperature value, does not present any significant drift, but only a random spread confined into a band of 0.6 mm.

In order to obtain more information and to validate the previous test results, 24 hours acquisitions, have been carried out at 1 Hz sampling frequency, with and without the external fan. The results (moving average) are presented in Fig. 4.6.

Without the external fan, the system presents a 4 h transient, characterized by an oscillating trend, due to an alternating operation of the internal fan. After this phase, the device temperature reaches a higher value that inhibits the shut-down of the fan,

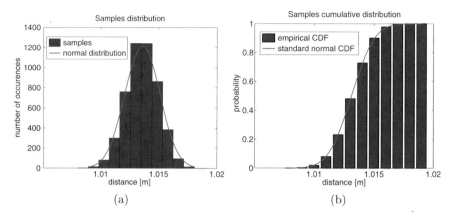

Fig. 4.7 Histogram (**a**) and cumulative (**b**) probability distribution functions of 5000 range measurements of the central pixel performed at 1000 mm from a planar wall

which continuously rotates and stabilizes the system until the end of the test. The second test confirms the hypothesis that the constant rotation of the fan can ensure a rapid stabilization of the measurement and a reduction of the transitory phase. In all the tests successively presented, a fan will be used and maintained in constant rotation, fixed to the rear of the Kinect V2™.

Finally, 5000 depth samples of the central pixel have been acquired with the same setup of the previous test. The histogram of the range measurements and the one of the cumulative probability distribution function are depicted in Fig. 4.7. It can be noticed that the sample has a Gaussian distribution. For this reason, in the following part of the article, all the datasets can be considered normally distributed in order to compute mean values and standard deviations.

4.3   Pixel-Wise Characterization

In this section, we analyze the range measurement at a single-pixel level. Uncertainty is defined according to the GUM framework as a precision corresponding to a casual or random error and an accuracy corresponding to a systematic error or bias (BIPM et al. 2008). Also, we investigate the uncertainty trend in function of the incident angle of the light on the target, as well as target intrinsic characteristics such as colors and materials.

Fig. 4.8 Anthropomorphic
robot setup for ground truth
motion

4.3.1   Setup

The setup is composed of a anthropomorphic robot arm ABB IRB 1600, characterized by a 0.02 mm repeatability in position. On the end effector there is an **HALCON** calibration pattern printed on a rigid aluminum plane, that helps the registration of the robot pose with the IR camera of the Kinect V2^{TM} that determines the Kinect V2^{TM} Spatial Reference System (SRS). Figure 4.8 shows the extrinsic calibration process that register the robot tool with the Kinect V2^{TM} SRS.

Kinect V2^{TM} is positioned on a photographic tripod and aligned with the planar white target mounted on the anthropomorphic robot end effector. Note that the robotic arm has a limited extension around 1000 mm; in order to cover an operating distance from 800 mm to 4200 mm, the range is divided in four parts with a 200 mm overlap. The four ranges are 800–1800 mm, 1600–2600 mm, 2400–3400 mm and 3200–4200 mm. In each of the range, the plane is translated with 20 mm steps and 4000 depth samples are acquired per pose. Note that for the uncertainty estimation along the sensor frame, the Kinect V2^{TM} is placed in the robot end effector and a planar wall is acquired. Planarity of such wall was smaller that the expected uncertainty for the Kinect V2^{TM} device.

4.3.2   Random Component of the Uncertainty in Space

In the first part of this test, the precision of the camera at different distances is evaluated. The standard deviation, representing the random component of the uncertainty, is plotted in Fig. 4.9 over the mean distance value. This test denotes

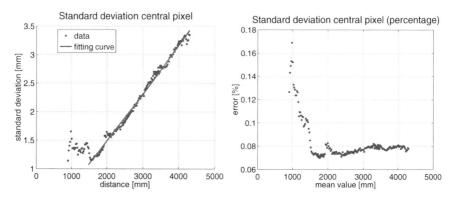

Fig. 4.9 Absolute and relative casual error in function of the distance

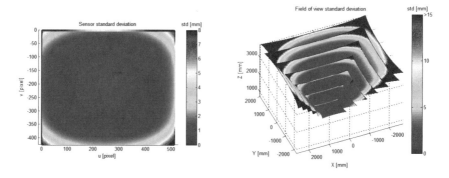

Fig. 4.10 Casual error in function of the pixel position and along the field of view

a linear trend ($R^2 = 0.9916$) from 1.2 mm at about 1500 mm to 3.3 mm at the maximum reliable distance (4200 mm) with an initial noisy baseline. This trend is attributed to a reduction of the quantity of light caught by each pixel, due to the increase of the target distance and is responsible for the degradation of the signal-to-noise ratio. Similar results can be found in Butkiewicz (2014).

Considering a set of Kinect V2™ acquisitions grabbed at 1250 mm far from a planar wall, the standard deviation of each pixel is calculated and plotted into a 424×512 image. Figure 4.10 shows how the random component of the uncertainty in the image and a trend can be extrapolated, the farther from the optical center, the bigger the uncertainty. This error could be attributed to the IR light cone illuminating the scene that is not homogeneous (Piatti and Rinaudo 2012). These lead corner points to be more noisy than the central ones.

Also, we analyses seven acquisitions grabbed in a field of view extended from 750 mm to 3750 mm along the camera optical axis. The cone trend is confirmed, we can denote that the measurement precision reaches up to 15 mm at the corner, which are probably not illuminated enough, and the resulting measurement less reliable.

4.3.3 Bias Component of the Uncertainty in Space

The aim of this test is to determine the accuracy of the single pixel distance measurement, determining the systematic error of the measurement. The experiment is conducted using the same setup described in Sect. 4.3.1 and the same data acquired in Sect. 4.3.2.

In Fig. 4.11, the difference between expected and measured distances is plotted. Four graphs, one for each step, are aligned minimizing the distance of the overlapping points. This harmonic shaped trend, called *wiggling error* in literature (Instruments 2014; Rapp 2007), is affected to the non-idealities of the electronics and most precisely on the non exactly sinusoidal shape of the modulated signal, that contains odd harmonic components. Also, a fitting curve ($R^2 = 0.9915$), calculated as a sum of sinusoidal functions, is plotted over the measured points.

Moreover, analyzing the data set acquired at 1250 mm, the mean value of each pixel's depth is calculated and a reference best fitting surface is computed on the point cloud obtained processing the meant distance data with the intrinsics parameters and the lookup table extracted from the SDK (*coordinate mapper*). Figure 4.12 denotes a distance error confined into a band of less than 20 mm; central points present a small positive displacement, while corner ones have a more important negative one. Furthermore, a point with a high displacement value can be noticed near the center of the image, probably due to a defect into the sensor matrix.

From these first tests, it can be deduced that the precision and the accuracy of the measurements are strongly correlated with the quality of the signal that is transmitted to the photo-diode sensors. First of all, since the light signal is diffused on the scene, its power is reduced respect to a single spot TOF system and the quality of the signal, especially its Signal-to-Noise Ratio (SNR), declines. Also, due to the imperfect lens distortion model, the geometrical position estimation provides an atrophied 3D measurement, especially on the corner pixels which are subject to higher distortions. In order to avoid such low quality measurement, the image can be

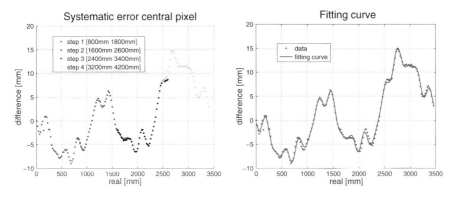

Fig. 4.11 Distance systematic error with fitting curve

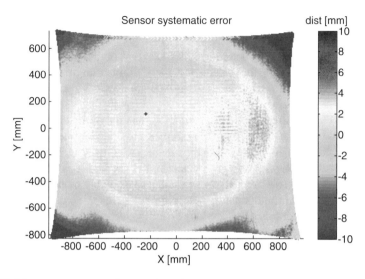

Fig. 4.12 Measurement systematic error along the sensor, at 1250 mm

cropped and only the central part of the depth image taken into consideration. Doing so, precision can be bounded by 2 mm and the accuracy by 4 mm at a 1250 mm range.

4.3.4   Error Due to the Incidence Angle on the Target

In this section, the influence of the incidence angle between the IR light signal and the target plane on the distance measurement is evaluated. The robotic setup guarantees a very accurate rotation around the sensor's **X**- and **Y**-axes.

In a first attempt, the central pixel behavior has been investigated. Four thousand distance frames measurements have been acquired with 5° step from the planar target parallel to the camera sensor to an inclination of 60°. The standard deviation of central pixel is calculated and depicted in Fig. 4.13, but no trend has been interpreted.

The test has been extended to a 62 × 130 pixels-wide central part of the sensor. With the help of a robotic arm, a planar white target is rotated into 13 poses around a vertical axis from 0° to 60° with 5° steps. The standard deviation of each pixel is calculated on 4000 depth measurement. The results are plotted in Fig. 4.14.

Analyzing the images, a band distribution (parallel to the rotation axis) of the computed standard deviation values is well evident and denotes its correlation with the incidence angle of the single ray. The higher is the angle, the lower is the amount of light per area unit returned by the target and so the SNR ratio. Still, precision is bounded by 1–1.8 mm. Similar results were obtained rotating around the horizontal axis.

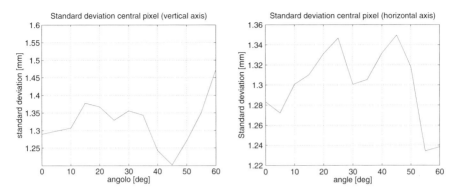

Fig. 4.13 Standard deviation value depending on target rotation angle around vertical (left) or horizontal (right) axis

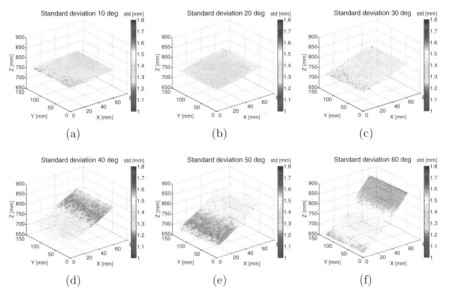

Fig. 4.14 Standard deviation related to the ray incidence angle. (**a**) 10° vertical. (**b**) 20° vertical. (**c**) 30° vertical. (**d**) 40° vertical. (**e**) 50° vertical. (**f**) 60° vertical

4.3.5   Error Due to the Target Characteristics

Here, we are investigating the influence of target color and material in the depth measurement. For this test, cardboard, plastic, aluminum, wood and fabric samples of black, white, yellow, red, blue and green colors have been used. We have attached the sample on a planar surface, one at a time. The camera was aligned at a distance of about 775 mm and 4000 acquisitions of depth values were performed.

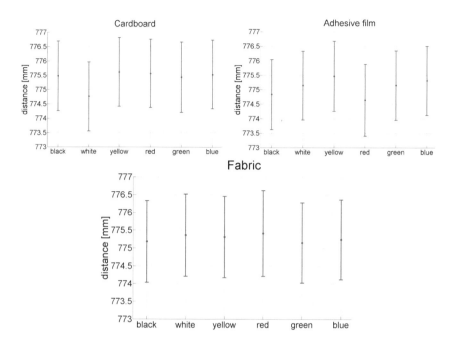

Fig. 4.15 Error distribution of different color targets for cardboard, adhesive and fabric film materials

4.3.5.1   Focus on Color

Figure 4.15 shows the results for the color on error bar graphs. A standard deviation value of about 1 mm is noticeable, constant in each set of acquisitions.

These tests denote an evident offset among the mean values of each set of acquired data. In particular, cardboard samples returned a 0.8 mm difference between the nearest and the farthest detected object. A similar error can be extracted from adhesive film data, but for different colors. For the fabric ones instead, the distance offset is confined into a 0.5 mm wide band.

Considering the sets of results together, the standard deviation, hence the precision of the measurement, does not depend on the color, nor the material, but is fixed at around 1 mm. However, the offset is impacted, hence the accuracy of the measurement. As far as the tests goes, such value does not seam to be correlated to the color.

4.3.5.2   Focus on Material

Dependency of the measurement respect to the target materials is discussed in this section. For this test, samples of cardboard, plastic, aluminum, wood and fabric materials were taken into consideration, regardless of its color. The test performed is similar to the previous one.

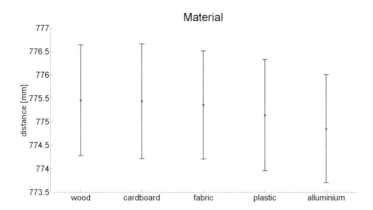

Fig. 4.16 Error distribution of different material targets

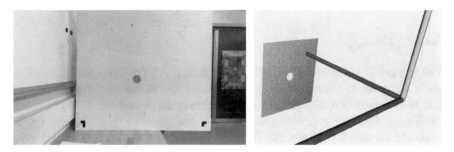

Fig. 4.17 Reflective tissue test: setup and corresponding point cloud

Figure 4.16 depicts the results of this test. It can be observed that the lower distance is associated with the most reflective material (aluminum). On the contrary, the object seen as the farthest by the camera is the wooden one, characterized by a rough surface. Also, in this test the standard deviation is almost constant and so no correlation between its value and the material can be pointed out.

4.3.5.3 Reflectivity

Finally, a circular sample of reflective tissue is acquired, a material commonly used in computer vision. Because of its high reflectivity, that saturates the camera pixels, this sample is not seen by the Kinect V2^{TM} at any distance. Figure 4.17 illustrates the setup where the reflective marker were put on the wood material. As a result, the Kinect V2^{TM} device does not return any depth measurement at the marker position.

Considering the previous analyses, it can be deduced that the color and the material of a target influence the depth measurement, but no trend has been pointed out. In concordance with Fankhauser et al. (2015), measurements are influenced by the reflectivity of the surface, since it indicates the quantity of light that bounces back to the sensor.

4.4   Sensor-Wise Characterization

In this section, we analyze the range measurement at a sensor scale. Still, uncertainty is defined according to the GUM framework (BIPM et al. 2008). Here, we investigate the capacity of the depth sensor to measure known geometry. We are also discussing on mixed pixel and multiple path errors.

4.4.1   Known Geometry Reconstructions

Here, the quality in the reconstruction of known geometries such as planes, cylinders and spheres is analyzed, measuring the distribution of the distances between the acquired points and the best fitting geometrical model.

4.4.1.1   Plane

The first geometric figure to be acquired is a plane, parallel to the sensor, disposed at about 1000 mm, in a setup similar than Sect. 4.3.1. At this distance, The point cloud has an extension of approximately 1400 mm along the **X**-axis and 1155 mm along the **Y**-axis. Also, the resolution along the direction identified by the **X**- and **Y**-axis is around 2.7 mm. It confirms the 70° × 60° field of view announced by Microsoft.

Within the acquired point cloud, a plane fitting operation is done, using a RANSAC fitting tool with refinement (Fischler and Bolles 1987). Figure 4.18a shows the acquisition and the fitting plane. Each points of the acquisition are painted in function of the distance from the model plane. A blue color represents a closed point, a green color an intermediate point and a red color a far point.

Figure 4.18b, c depict the error distribution into the sensor area. Each points of the acquisition are painted in function of the distance from the model plane. An orange color represents a closed point, while a green-blue color far point in a

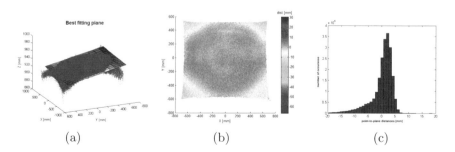

(a)   (b)   (c)

Fig. 4.18 Reconstruction of a plane. (**a**) Best fitting plane. (**b**) Point-to-plane distances. (**c**) Histogram of distances

direction and a red color in the opposite direction. The greater distortion, of around 15 mm, is observed at the corners of the image. Nevertheless, in the central part of the image, the distance is confined into a band of about ±5 mm, while, around the corners, it reaches values of tens of millimeter.

This error has been in part imputed to the transmitted light signal that has a conical diffusion shape. Light signals may not have enough power at the corners of the sensor respect to the environment light and then produce noisier measurements. Also, the IR camera lens introduces optical distortions that are not completely compensated by the mathematical model-based conversion from depth images into point clouds.

4.4.1.2 Cylinder

The cylinder is a simple object, commonly used to test the quality of 3D reconstruction (Russo et al. 2007). A cylindrical-shaped objects, with a 500 mm diameter was placed at a distance of about 1000 mm from the sensor. The acquisition is shown in Fig. 4.19a, as well as a best fitting cylinder obtained with a RANSAC fitting and successively refined.

Figure 4.19b, c depict the alignment results, highlighting the distance errors distribution; where most of the acquired data are less than 15 mm distant from the model. Furthermore, as stated in Sect. 4.4.1.1, the points with a greater deviation ($\geq$ 30 mm) are those corresponding to the upper and lower edge of the sensor, due to the lighting cone and the lens distortion.

In this test, it is interesting to see how the uncertainty in the border of the frames influence way more the uncertainty that the surface orientation. Actually, it seems that the orientation different from the orthogonal one do not influence the quality of the reconstruction. This hypothesis will be verified with the sphere.

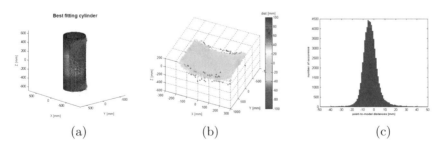

(a)   (b)   (c)

Fig. 4.19 Reconstruction of a cylinder. (**a**) Best fitting cylinder. (**b**) Point-to-model distances. (**c**) Histogram of distances

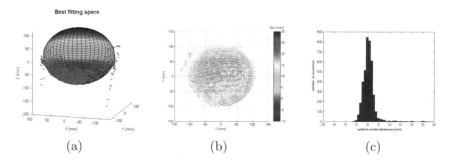

Fig. 4.20 Reconstruction of a sphere. (**a**) Best fitting sphere. (**b**) Point-to-model distances. (**c**) Histogram of distances

4.4.1.3   Sphere

Finally, the quality of the reconstruction of a spherical surface is evaluated, scanning a portion of a 105 mm-radius sphere-shaped object, 1000 mm far from the sensor. Figure 4.20a shows the point cloud acquisition the sphere model aligned with the RANSAC algorithm.

Figure 4.20b, c show the error distribution, almost symmetric, centered in zero and confined in a band of less than 5 mm.

From this test, it is possible to confirm the previous hypothesis for angular surface influence. Also, it is worth noting how contour points are tens of millimeters distant from the model surface. This must be due to mixed pixel error, described in the next section.

4.4.2   Mixed Pixels Error

This test aims to analyze the quality of edge reconstruction. It is known from the characteristics of the Kinect V2^{TM} that the depth map endures an edge-aware filter in order to remove mixed pixels error. A planar surface was posed about 350 mm far from a 1 m^2 flat background. Point clouds are acquired, with the Kinect V2^{TM} device manually aligned with the setup, at a distance of about 1050 mm from the target. Also, the background is acquired.

The acquisition denotes some spurious points in the jump from foreground to background, as shown in red in Fig. 4.21b. We used two RANSAC segmentation algorithms available in Point Cloud Library (PCL) for the identification of the two planes, as shown in Fig. 4.21a. The remaining points, not belonging to any plane are shown in red, corresponding to the mixed pixels.

This phenomenon, also called jump edge by Ratshidaho et al. (2014), is a consequence of a weighted average operation in the depth map shown by

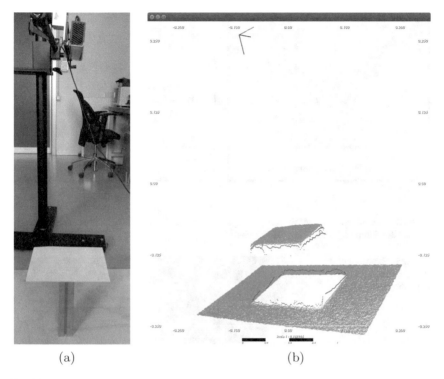

Fig. 4.21 Experimental setup (**a**) showing the acquired point cloud with mixed pixel enlighten in red (**b**). (**a**) Setup. (**b**) Point cloud acquisition

Instruments (2014), applied to neighboring pixels, in order to reduce the noise. As a consequence, an intermediate value is attributed to contiguous pixels belonging to this transition areas.

This test was extended to a situation in which the background plane is too far to be acquired (depth ≥ 4500 mm). In this case, the mixed pixel error is absent.

4.4.3 Multiple Path Error

TOF cameras have a well-known problem called multiple path error. This phenomenon originates from the multiple reflections generated into concave geometries in which part of IR light is reflected from one surface to the other, before returning to the camera sensor (Dorrington et al. 2011; Bhandari et al. 2014).

The signal covers longer paths than the expected ones and in particular the emitted light produces noise for the surrounding pixel. The light received by a given pixel is not only the shortest path but a combination of multiple signals that are

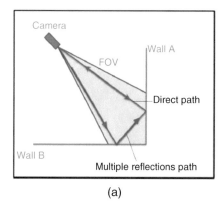
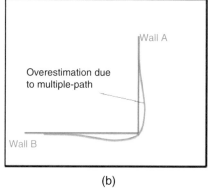

(a)   (b)

Fig. 4.22 This schemes represent how the light signal bounces on multiple concave surfaces (**a**) and how the depth acquisition is overestimated (**b**)

Fig. 4.23 Experimental setup for multiple path uncertainty estimation

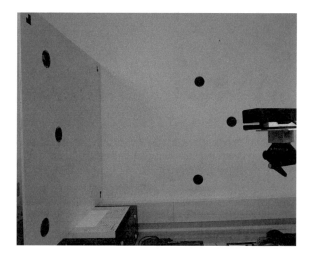

interpreted as a mean value in between. Figure 4.22a illustrates the combination of the direct path with the multiple refections ones, while Fig. 4.22b shows the mean effect that overestimates the depth TOF measurement.

In order to analyze the multiple path error, a specific setup is built with two surfaces (a white wall and a wooden panel), opportunely aligned, in order to create a dihedral and to intensify such error. The experimental setup is presented in Fig. 4.23.

A ground truth acquisition has been obtained with a Leica TS06 Total Station. A set of three points on each of the surfaces has been accurately acquired interpolated with best fitting planes and took as reference measurement. Note that the Total Station also uses a TOF technology, but since it focuses the signal along a single laser bean, it does not suffer the multiple path reflection.

Since the two systems were in different positions, a registration is necessary to align the Kinect V2^{TM} acquisition with the interpolated planes. To do so, three black circular markers were placed on both the surface, as seen in Fig. 4.23. Knowing the 3D coordinates of those points respect to the acquisition reference system, it is possible to associate the corresponding point and align the two reference systems.

In particular, with the Total Station, the points were directly measured. For the Kinect V2^{TM} measurement, finding the barycenter of the black circular markers on the IR frames permits their projection in 3D. In order to reduce the multiple path effect on the planar estimation, those markers were placed as far as possible to the corner.

Successively, the minimization of the distance between the corresponding marker was done with an Singular Value Decomposition (SVD) based registration pipeline in order to find the extrinsic registration parameter between the two system SRS.

The Fig. 4.24 depicts the alignment error between the point cloud and the reference planes. An error of some tens of millimeter (up to 80 mm) affects the pixels less than 250 mm distant from the plane-plane intersection line. The multiple path reflection is the major technological issue of TOF cameras.

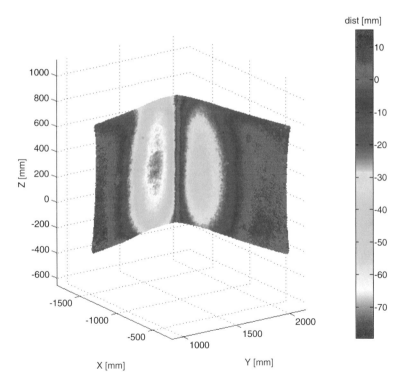

Fig. 4.24 Multiple path uncertainty estimation obtained from the distance to the walls

4.5 Conclusion

In this chapter, the uncertainty in depth measurement for the commercially available Microsoft Kinect V2^{TM}, considered as the state-of-the-art for TOF cameras has been characterized following the GUM. First of all, the TOF signal has been investigated, questioning the reliability of the three main signal frequencies. Also, depth measurements are sensitive to temperature and an extensive usage induces significant drifts in depth measurements reaching up to 6 mm.

Successively, a pixel-wise investigation focused on the normal distribution of the depth uncertainty in space. The random component of the uncertainty in depth has been characterized, such precision increases linearly with the sensor distance, bounded by 1.5 mm at a 2 m range and reaching 3.5 mm at a 4.5 m range. The random uncertainty has a cone shape in 3D space, due to the illumination system that enlightens the scene with its Light Detection And Ranging (LiDAR) signal. In the corner of the matrix sensor, where only a small amount of light projects, the random component error can reach up to 15 mm. Further studies showed a *wiggling error* for the biased component of the uncertainty in depth, bounded in ±15 mm. This bias is accentuated on the corner and can reach up to tens of millimeters, but can be compensated knowing the model.

Extended tests show up the capacity of the Kinect V2^{TM} to measure simple shapes in a surrounding environment. The most important component of the systematic uncertainty actually comes from the mixed pixels effect and the multiple path reflections, that create biases reaching up to 80 mm. Such bias is the main source of uncertainty in Time-of-Flight cameras and cannot be compensated; scenes to reconstruct have to be chosen with care, since any concave geometry can create multiple path reflections.

Chapter 5
Metrological Qualification of the Orbbec Astra S^{TM} Structured-Light Camera

The Orbbec Astra S^{TM} is a promising Three-dimension (3D) camera based on the Structured-Light technology. The Chinese company released their sensor in 2015, based on a proprietary Infra-Red (IR) depth sensor. Whether it is not clear what is the physical resolution of the depth sensor, the camera can produce point cloud up to 1280×1024 at 5 Hz (Table 5.1). In this chapter, we limit the resolution to 640×480, to maintain a fair comparison with a 30 Hz frame rate. The baseline between the camera and the project is equal to 75 mm. Even though the range is limited to 2 m, the device provide a point cloud up to 2.5 m, hence we characterized it up to that range. In this chapter, we first focuses on preliminary test including the calibration of the device. Then, we provide an estimation of the random and systematic error components, in a similar fashion than previously obtained with the Kinect V2^{TM}. Last, we provide qualitative results for simple shape reconstruction.

5.1 Preliminary Experiments

As a fist step, we present the process of calibration of both cameras of the Orbbec Astra S^{TM}. We use the default setup with a resolution of 640×480 for both. For each camera, we captured 15 images of the calibrator, as shown in Fig. 5.1 for the IR camera. With the HALCON software, we estimate the intrinsic parameters shown in Table 5.2, with a reprojection error equal to 0.19 pixel. Repeating the setup four times, we show that the radial distortion are negligible as well as the rectangularity and skewness of the pixels. We believe the cameras were calibrated in factory. In addition to the intrinsic parameters, we estimate the extrinsic parameters by grabbing images with both cameras simultaneously. The color and the depth Spatial Reference Systems (SRSs) are distant by roughly 25 mm, accurate results are shown in Table 5.2.

© The Author(s), under exclusive licence to Springer International Publishing AG, part of Springer Nature 2018
S. Giancola et al., *A Survey on 3D Cameras: Metrological Comparison of Time-of-Flight, Structured-Light and Active Stereoscopy Technologies*, SpringerBriefs in Computer Science, https://doi.org/10.1007/978-3-319-91761-0_5

61

Table 5.1 Orbbec Astra S^{TM}: main characteristics

IR camera resolution	640 × 480	(pix)
RGB camera resolution	640 × 480	(pix)
Maximum frame rate	30	(Hz)
Field of View (FOV)	60(H) × 49.5(V)	(°)
Measurement range	400–2000	(mm)
Baseline IR projector	75	(mm)
Dimension	160 × 30 × 40	(mm)
Weight	300	(g)
Connection	USB 2.0	
Operating system	Windows 8/10, Linux, Android	
Software	Astra SDK, OpenNI2, 3rd party SDK	

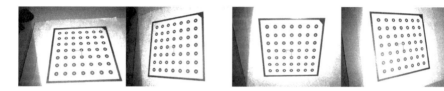

Fig. 5.1 Example of images used for the calibration of the IR camera

Table 5.2 Orbbec Astra S^{TM}: intrinsic parameters for both cameras

Parameter		IR camera		RGB camera	
		Nominal	Estimated	Nominal	Estimated
Focal length	f	2.98 mm	3.04 mm	1.98 mm	1.96 mm
Optical center X	c_x	320 pixel	313.3 pixel	320 pixel	316.3 pixel
Optical center Y	c_y	240 pixel	241.3 pixel	240 pixel	238.3 pixel
Translation X	t_x	–	–	−25 mm	−25.48 mm
Translation Y	t_y	–	–	0 mm	−0.43 mm
Translation Z	t_z	–	–	0 mm	−0.64 mm
Rotation X	r_x	–	–	0°	−0.8°
Rotation Y	r_y	–	–	0°	0.4°
Rotation Z	r_z	–	–	0°	−0.1°

The first shrewdness for the Orbbec Astra S^{TM} characteristics assessment regards the spatial resolution, especially in depth. Figure 5.2 shows the top view of a point cloud acquired with the Orbbec Astra S^{TM}. Sampling along **X**- and **Y**-axes is expected due to the sensor resolution, but depth quantization is not.

Figure 5.3 shows the spatial resolution trend over depth measurement, along the **X**-, **Y**- and **Z**-axes. Such data have been assessed acquiring point at different ranges and estimating their resolution over depth and **X**- and **Y**-sensor axes. Resolution along **X**- and **Y**-sensor axes are increasing linearly while the depth one along **Z**-grows up exponentially.

Analogous results are reported by Smisek et al. (2013) with the Kinect V1^{TM} structured-light depth camera. Such quantization variation has been interpreted

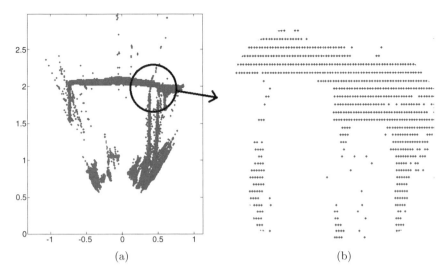

Fig. 5.2 Spatial resolution shown on a point cloud. (**a**) Acquired point cloud (**XZ** plane). (**b**) Zoom in highlighting depth resolution

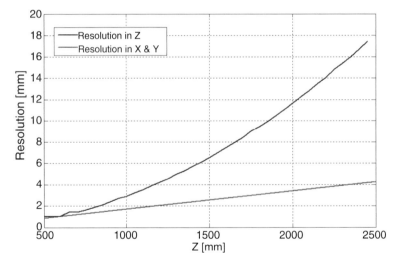

Fig. 5.3 Spatial resolution over depth

as the effect of the intersection of the camera resolution with the projector one. Figure 5.4 illustrates such effect, showing how the depth measurement occurs in structured-light system. The depth quantization along the optical axis increases exponentially. Nevertheless, such model does not perfectly agree with the experimental results, since the quantization should be constant along the **X**- and the **Y**- sensor axes while it seems to be nodding.

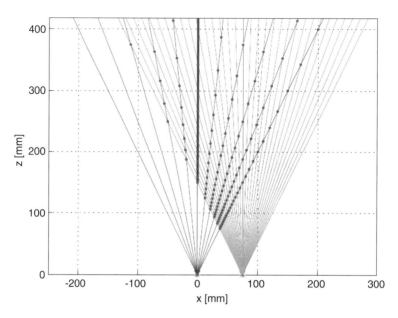

Fig. 5.4 Spatial resolution interpretation

5.2 Random Error Component Estimation

The random error of the Orbbec Astra S^TM is estimated with a setup similar than in the Kinect V2^TM tests (see Fig. 4.8). We use an anthropomorphic robot with a moving target on the end effector, the which position is controlled with a 0.02 mm repeatability. The structured-light device measures the depth of a target placed orthogonally of the camera optical axis. Due to the limited range of the robotic arm, the test has been repeated for four different ranges in order to cover a 500–2500 mm range. Note that the target planarity and orthogonality uncertainties are negligible respect to the expected Orbbec Astra S^TM uncertainty and such target is covered with a non-reflective white material in order to prevent excessive reflections.

The random component of the error along the optical axis is shown in Fig. 5.5. The quantization effect, previously described, impacts drastically on the normality of the uncertainty. The standard deviation is scattered, while the dashed line represents the *type B* uncertainty due to the resolution. Since the *type A* stochastic uncertainty is less than the one introduced by the resolution, the Guide to the Expression of Uncertainty in Measurement (GUM) (BIPM et al. 2008) recommend to consider the *type B* uncertainty for the random component of the measurement. Such uncertainty is equal to $\frac{R}{2\sqrt{3}}$, R being the resolution.

The random component of the uncertainty along the sensor plane is shown in Fig. 5.6. To do so, the Orbbec Astra S^TM device has been mounted on the robot

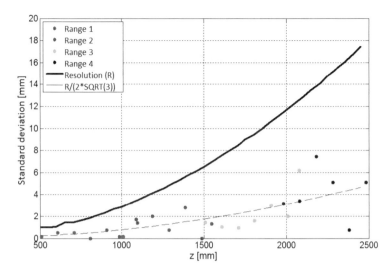

Fig. 5.5 Random error component in depth (central pixel)

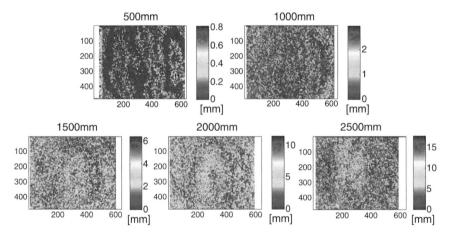

Fig. 5.6 Random error component distribution in sensor

and a planar wall has been used as target. The wall planarity is considered to be more accurate than the expected uncertainty of the Orbbec Astra S^{TM} device. The scattered distribution of the uncertainty supports the *type B* evaluation hypothesis. Uncertainty in depth measurements is bounded by a ± 15 mm at a 2500 mm range. In comparison, Kinect V2^{TM} reaches similar uncertainty on the border of the sensor (see Fig. 4.10), but the central part enlightened by the light cone stays below a ± 5 mm range.

5.3   Systematic Error Component Estimation

The systematic error components along the optical axis and the sensor plane are estimated with similar setups. The systematic error component of the central pixel in function of the depth is represented in Fig. 5.7 and increases quadratically with the depth. Such bias can reach up to 100 mm at a 2500 mm range with the Orbbec Astra S^{TM} device.

The systematic error component along the Orbbec Astra S^{TM} sensor (Fig. 5.8) is similar to the results obtained for the Time-of-Flight (TOF) system (Fig. 4.12). The farther from the optical center, the most uncertain the depth measurement. This is mainly due to an non-perfect estimation of the radial distortions. Nevertheless, the bias effect is increasing quadratically with depth reaching up to 100 mm. Also, interferences occur along the **X**-axis of the sensor due to the lateral position of the projector respect to the IR camera sensor.

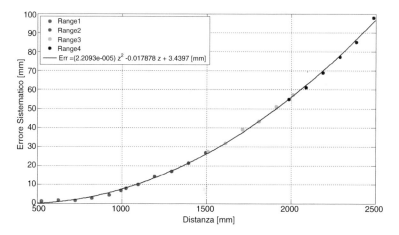

Fig. 5.7 Systematic error component in depth (central pixel)

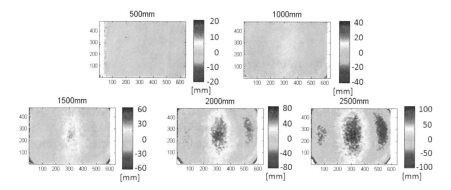

Fig. 5.8 Systematic error component distribution in sensor

5.4   Shape Reconstruction

In this part, the reconstruction of simple geometrical shapes are evaluated, in a similar way that it was done with the Kinect V2^{TM}. Distribution of distances between point clouds and simple shape models is evaluated, for the sphere, the cylinder and a convex corner shape. The models are aligned using a RANSAC registration pipeline refined by an ICP method (Besl and McKay 1992).

5.4.1   Sphere

A white sphere with a 400 mm diameter has been placed at a 800 mm range from the sensor. Results are illustrated in Fig. 5.9, with colors evidencing the deviation from the model. Annular shaped artifacts are noticeable at different depths, identified by a scalar deviation, showing the depth quantization. Also, the borders are jagged due to the orientation of the target surface. The error oscillates around ±2 mm, which actually looks slightly better than the Kinect V2^{TM}, but for a smaller sphere and at a larger distance (see Fig. 4.20c). Nevertheless, such deviation distribution qualifies the random component of the uncertainty and may be biased.

5.4.2   Cylinder

Another simple shape reconstruction is performed on a white cylinder with a 240 mm diameter and a 780 mm height. Results are shown on Fig. 5.10, with the cylinder placed at a 700 mm distance. The quantization artifact is once again visible

Fig. 5.9   Sphere: point cloud acquisition and sphere fitting using an Orbbec Astra S^{TM} device

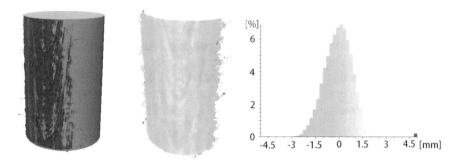

Fig. 5.10 Cylinder: point cloud acquisition and cylinder fitting using an Orbbec Astra S^{TM} device

Fig. 5.11 Concave corner: point cloud acquisition and plane fitting using an Orbbec Astra S^{TM} device

on the depth measurement, as well as noisy point on the border of the cylinder. The error distance between the point cloud and the fitting model is coherent with the expected precision of the sensor. For the sphere and cylinder fitting shape measurement, the Gaussian normality hypothesis is confirmed.

5.4.2.1 Concave Corner Shape

The last simple shape reconstruction consists in a concave corner shape, where TOF system actually provides a worse depth biases due to multiple path reflections. The acquisition of the corner, composed of two planar walls, has been performed at a 1 m distance. Figure 5.11 shows the reconstruction as well as the deviation around the two planar models. Results show coherent error on the reconstruction, that does not seem to be affected by the wall reflection. Deviation ranges around ±4 mm.

5.5   Conclusion

In this chapter, we have shown how a state-of-the-art Structured-Light vision system performs on similar test realized on the Kinect V2^{TM} TOF device. For this purpose, The Orbbec Astra S^{TM} Structured-Light camera has been used. Results shows that the uncertainty as well as the bias for the Orbbec Astra S^{TM} camera measurements increase quadratically with depth. In comparison, Kinect V2^{TM} uncertainty is linear and bias is bound by a ± 15 mm *wiggling error*. Uncertainty distribution along the sensor for the Orbbec Astra S^{TM} is casual and bias increases in corners. In comparison, Kinect V2^{TM} perform better in the area enlightened by the IR modulated signal. Table 6.2 summarizes the depth performance at 1000 and 2500 mm depth. Also, simple shape reconstruction showed similar results for Orbbec Astra S^{TM} and Kinect V2^{TM} devices. The only inconvenience for the Kinect V2^{TM} device is the multiple path reflection, typical drawback for TOF cameras, which is the most important error component.

Accuracy and precision in depth is evaluated for the Kinect V2^{TM} device under multiple circumstances and successively compared with a Structured-Light depth camera. Unlike TOF technology, Structured-Light systems provide better accuracy but worse precision. Nevertheless, multiple measurements can improve the random component uncertainty of the Structured-Light system, but the bias component of the TOF depth measurement due to multiple path reflections can hardly be compensated.

Chapter 6
Metrological Qualification of the Intel D400^{TM} Active Stereoscopy Cameras

In January 2018, Intel released the RealSense RS400^{TM} generation of Three-dimension (3D) computer vision devices based on Active Stereoscopy technology. It includes RGB-D modules and image processing hardware, providing makers and system integrators with a flexible way to implement embedded 3D vision solutions. Currently, Intel's RS400^{TM} generation presents state-of-the-art specifications in terms of resolution and operating range (Table 6.1), considering commercial mass market 3D active stereoscopy cameras. Intel propose two different vision modules, the RS410^{TM} and the RS430^{TM}, based on the D4 processor that the they implement in two ready-to-use cameras, the D415^{TM} and the D435^{TM}. As for the Kinect V2^{TM} and the Orbbec Astra S^{TM}, we are investigating the metrological performances of the Intel D415^{TM} and D435^{TM} depth sensor. First, we present a preliminary analysis of the devices. Then, we focus our test on pixel-wise characterization along the depth, the sensor plane and the angular parameters. Last, performances for shapes reconstruction are analyzed.

6.1 Preliminary Experiments

The D415^{TM} and the D435^{TM} estimate depth with a structured-light technology but with two different intent. The D415^{TM} is intended for finer reconstructions in a narrow field-of-view, whereas the D435^{TM} is intended for acquisitions over a much wider field-of-view, but with a worst spatial resolution. This stands out immediately from the cameras' technical data: the D415^{TM} has a larger baseline (about 55 mm vs 50 mm) and a greater focal length than the D435^{TM}. Also, it is worth noting that the D435^{TM} has a global shutter sensor, whereas the D415^{TM} has a rolling-shutter one, hence D415^{TM} is expected to be more accurate when dealing with static scenes, but would not perform well in highly dynamic scene.

S. Giancola et al., *A Survey on 3D Cameras: Metrological Comparison of Time-of-Flight, Structured-Light and Active Stereoscopy Technologies*, SpringerBriefs in Computer Science, https://doi.org/10.1007/978-3-319-91761-0_6

Table 6.1 D400^{TM} cameras: main characteristics

	D415^{TM}	D435^{TM}	
IR camera resolution	1280 × 720	1280 × 720	(pix)
RGB camera resolution	1920 × 180	1280 × 720	(pix)
Maximum frame rate	90	90	(Hz)
Baseline	55	50	(mm)
Field of View (FOV)	69.4(H) × 42.5(V)	91.2(H) × 65.5(V)	(°)
Measurement range	160–10,000	200–4500	(mm)
Dimension	99 × 20 × 23	90 × 25 × 25	(mm)
Weight	60	100	(g)
Connection	USB 3.0		
Operating system	Windows 8/10, Linux		
Software	RealSense SDK, hand and face tracking		

Fig. 6.1 D415^{TM}'s pattern (on the left) and D435^{TM}'s pattern (on the right)

The two cameras show very different Infra-Red (IR) patterns, as shown in Fig. 6.1. The D435^{TM} has a random focused dot pattern, while the D415^{TM} has a more regular pattern that replicates. The projector of both camera can be switched off and its intensity can be tuned manually according to the environment lighting conditions.

The scope of the preliminary tests is to qualitatively understand the cameras' behavior and the effects operating parameters tuning's on the depth estimation. The cameras is factory-calibrated and the intrinsic and extrinsic parameters of the sensors are stored on board, easily accessible via the *librealsense* APIs. Considering the focal lengths and the baseline between IR sensors, one can compute the theoretical random error in space due to pixels spatial quantization. Figure 6.2 plots the theoretical uncertainty in function of the depth, based on the resolution along **X**-, **Y**- and **Z**-axes. The D415^{TM} has better theoretical performances, the spatial resolution along the **Z**-axis has been computed assuming a unity disparity error. Keselman et al. (2017) states that the actual disparity error is generally about 0.1 pixel for active systems. As a further consideration, the depth reference frame is the same as the left infrared camera.

Depth estimation is performed in-hardware by the D4 processor, the *librealsense* APIs allow the user tuning algorithm's parameters so as to optimize the result on specific applications. To facilitate the tuning, Intel also propose different *"Visual*

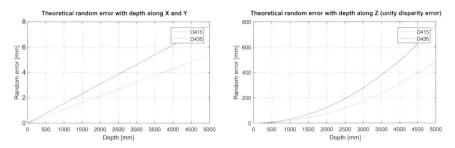

Fig. 6.2 Theoretical uncertainty for D400^{TM} cameras

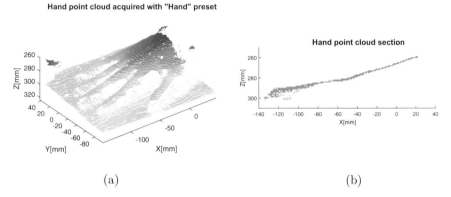

(a)   (b)

Fig. 6.3 D415^{TM}'s "Hand" preset acquisition. (**a**) Hand point cloud. (**b**) Point cloud X-Z cross-section

presets", they are settings profiles for specific purposes. Such profiles are saved in parameter files (.json format) that are loaded via the APIs. As tuning each parameter in an optimal way is a challenging task, Intel provides *Visual presets* for typical use cases: "Default", "High accuracy", "Hand", "High density", "Medium density". The presets are embedded in the APIs, details about each one are on the *librealsense*'s github repository and the RealSense community provides and share custom ones. Figure 6.3 shows a point cloud of a hand captured with the "Hand" visual presets, which proves to be optimized for such a scene. On the right side, a cross-section on the **X Z** plane is depicted, here the quantization effect along depth can be noticed.

To test the metrological properties of the presets, stability and spreading in measuring a constant depth value of the central pixel is verified capturing a fixed target over about 8.5 min with the "Default" and the "High accuracy" presets. Results show that "High accuracy" performs slightly better both in stability and measure variance. The filtering in the "High accuracy" preset distort the distribution from a Gaussian, with a negligible improvement in the variance of the measurement.

Nevertheless, they differ in the depth estimation around objects edges. The map produced with the "High accuracy" shows many holes, particularly around edges

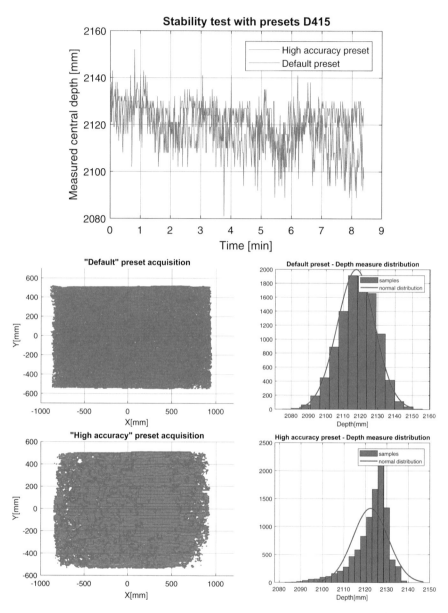

Fig. 6.4 *Top:* stability test with "Default" and "High accuracy" presets for the D415^{TM}. *Bottom:* Acquisition and distribution of the depth measurements with both presets

(where measurement would be less accurate), compared to the one produced with the "Default" presets (Fig. 6.4).

Intel D400^{TM} cameras are passively cooled, thus thermal stability over long operating times are investigated. The cameras are turned up for about 17 h measuring

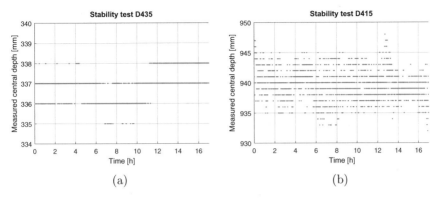

Fig. 6.5 Stability test results. (**a**) D435^{TM}. (**b**) D415^{TM}

a static target with the central pixel. Both devices have not shown any particular deviation during the operating time, such a result is intuitive to understand as the depth estimation does not rely on semiconductors' physical principles (like the Kinect V2^{TM}), but on numerical algorithms that are not directly affected by thermal effects. Also, the quantization is shown to be constant at different ranges and the variation of the measurements is shown to increase with the depth (Fig. 6.5).

6.2   Pixel-Wise Characterization

Here, we investigate the random and the systematic component of the uncertainty at a pixel level. The uncertainty is investigate both along the depth and along the sensor plane. In this section, we only relate the uncertainty for the D415^{TM}.

This test are realized in a fashion similar than the for the Kinect V2^{TM} and for the Orbbec Astra S^{TM}. For this test, we rely on a set of two robots coordinated between each others, as shown in Fig. 6.6. The D415^{TM} was fixed on the end effector of one robot and a target is moved by the second one. The advantage of using such setup is to include the camera depth range within the robot range and realize a single experiment instead of four like for the Kinect V2^{TM} and the Orbbec Astra S^{TM}. The robot used for moving the target is a *Fanuc R-2000iB*, ensuring a repeatability of 0.15 mm.

The calibration plate from **HALCON** is mounted on the target to perform the camera-target alignment, in order to calibrate and align the robot end-effector with the camera Spatial Reference System (SRS). After the calibration, the target is covered with a white surface and moved step-by-step along the **Z**-axis of the robot tool from around 0.7 m to around 2.6 m.

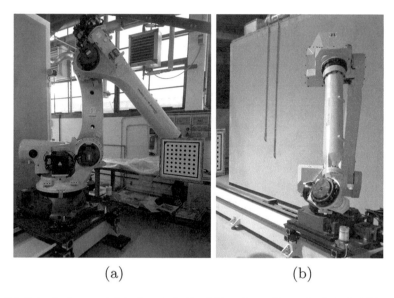

(a)   (b)

Fig. 6.6 Setup for the uncertainty characterization (**a**) target mounting on the first robot (**b**) camera mounting on the second robot

6.2.1 Random Component of the Uncertainty in Space

The random component of the error is evaluated along depth. We perform 3000 acquisitions of the central pixel at different depths and report the standard deviation in Fig. 6.7. We show the random error component as well as its relative value. The random error is bounded by 2 mm for distances below 1100 mm. The random error grows quadratically with the depth, as expected from the theoretical depth resolution of the D415^{TM}. We can see that it is bounded quadratically. However, we denote a discrepancy between the random error and its upper bound. We argue that this phenomenon is due to the depth triangulation, in particular the fact that some point are triangulated while other are just spatially interpolated.

Successively, the random component of the uncertainty is evaluated along the camera plane directions. In a similar setup, we acquire 3000 depth maps of a flat wall at about 0.5 m, 1.0 m, 1.5 m and 1.9 m. The standard deviation will provide an estimate the random component of the uncertainty along the camera sensor. Figure 6.8 show that the random component is bounded by 2.21 mm at 0.5 m, 12.86 mm at 1.0 m, 32.99 mm at 1.5 m and 85.56 mm at 1.9 m. Also, by analysing the shape of the random component of the uncertainty along the sensor, we can identify a regular pattern. We believe it is due to the pattern projected by the Structured-Light technology.

Finally, Fig. 6.9 provide an overview of the random component of the uncertainty in 3D space.

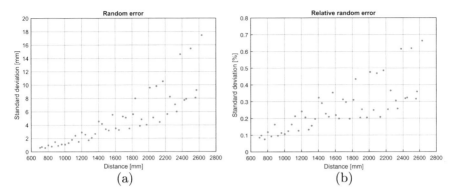

Fig. 6.7 D415^{TM}'s central pixel's standard deviation with depth. (**a**) Absolute random error. (**b**) Relative random error

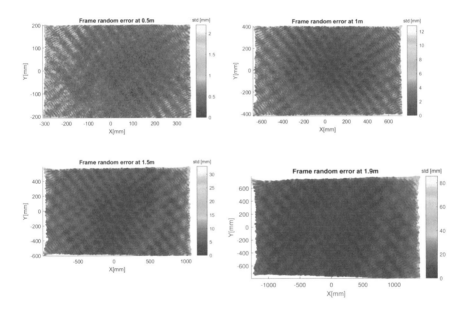

Fig. 6.8 D415^{TM}'s frame systematic error with depth

6.2.2   Bias Component of the Uncertainty in Space

The systematic component of the uncertainty is evaluated as the difference between the displacement measured by the camera at each step and the reference one imposed by the robot arm. The comparison is performed over the same 3000 samples previously used for the random component, between the average value

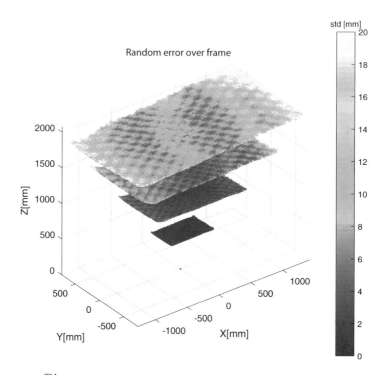

Fig. 6.9 D415^{TM}'s frame random error cone with depth

of the central pixel and the robot displacement. The bias is bounded by 5 mm for distances below 1100 mm and grow quadratically. For the similar technological motivation, we believe that the bias oscillation originates from the projected pattern (Fig. 6.10).

The bias component over the frame is determined as each pixel's distance from the interpolating plane. We produce the point cloud from the depth map and interpolate the acquisition with a plane. The test is performed on the same data used for the random component's one along the sensor plane. Figure 6.11 show that the systematic component of the uncertainty is bounded by 9.164 mm at 0.5 m, 26.75 mm at 1.0 m, 61.7 mm at 1.5 m and 107.24 mm at 1.9 m. Analysing the systematic component of the uncertainty along the frame, we can identify an hallo centered in the center of the frame. Such effect was already presented for the Orbbec Astra S and correspond to the non-linearity introduced by the lenses.

Finally, Fig. 6.12 provide an overview of the systematic component of the uncertainty in 3D space.

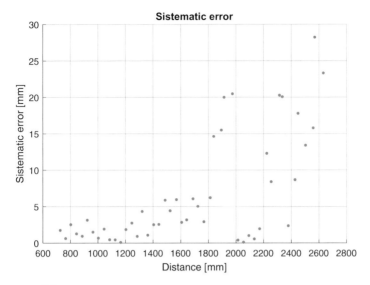

Fig. 6.10 D415^{TM}'s central pixel systematic error with depth

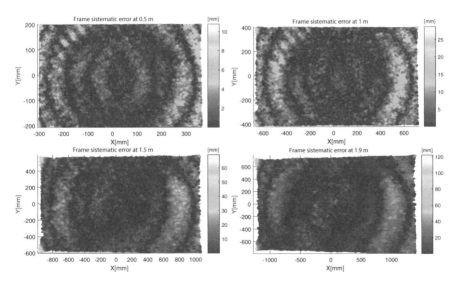

Fig. 6.11 D415^{TM}'s frame systematic error with depth

6.2.3 Uncertainty Due to the Orientated Surfaces

Last, the effect of the orientation of the surface on the depth uncertainty is evaluated. We orient our target with 5° rotation steps around both vertical and horizontal axis. Figure 6.13 shows the standard deviation of the central pixel's measurement in function of the orientation.

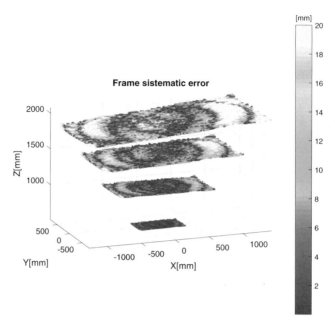

Fig. 6.12 D415^{TM}'s frame systematic error cone with depth

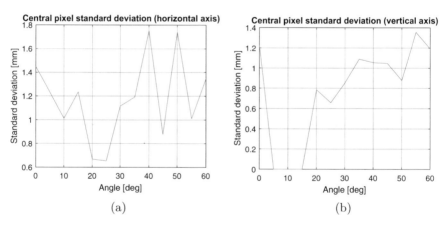

Fig. 6.13 D415^{TM}'s central pixel's std deviation with facing angle. (**a**) Horizontal axis. (**b**) Vertical axis

Figure 6.14 shows the standard deviation over the target surface at different poses (around the vertical axis). Note that the lower the angle, the better the shape reconstruction of the point cloud. Moreover the standard deviation is randomly distributed over the target surface, as expected by a the Active Stereoscopy technology.

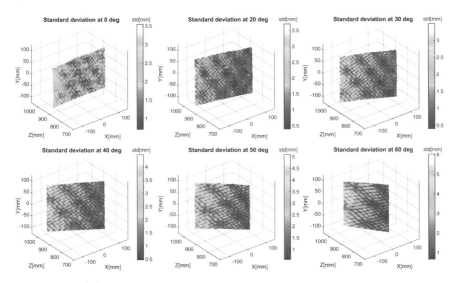

Fig. 6.14 D415^{TM} frame random error with facing angle (0–60°)

6.3 Sensor-Wise Characterization

Those test aims to characterize the overall depth reconstruction performances while reconstructing geometrical shapes such as plane, cylinder and sphere. We also verify with the last test that a sharp edge is reconstructed without any mixed-pixel effect due to an eventual averaging operation to smoothen the depth.

6.3.1 Plane Reconstruction

The point cloud of a planar surface is acquired and successively fitted over a plane model to estimate the cloud-model distance. Results are shown in Fig. 6.15. For what concerns the two cameras, the D415^{TM} shows a symmetrical modulation of the cloud due to the lenses distortion. In contract, while the D435^{TM} does not show any significant effect. The distance distributions for the two cameras denote quite similar performances, even though the D435^{TM}'s distribution has a slightly more Gaussian trend.

6.3.2 Cylinder Reconstruction

We perform similar test with a cylindrical shape of diameter 500 mm. The acquired point cloud is registered via an Iterative Closest Point (ICP) procedure to the 3D

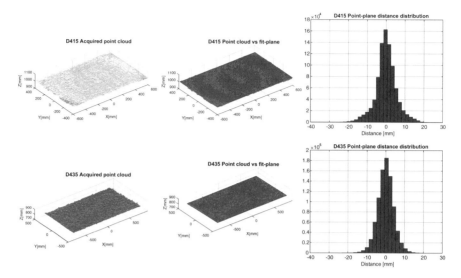

Fig. 6.15 Acquired point cloud, fitting plane and distance distribution. **Top:** results for the D415^{TM}. **Bottom:** results for the D435

model and then the cloud-model distance is computed. The qualitative analysis of the point cloud smoothness and the distance distributions narrowness in Fig. 6.16 shows that the D435^{TM} generates a noisier point cloud than the D415^{TM}.

6.3.3   Sphere Reconstruction

Similar test are performed with a spherical object of radius equal to 120 mm. Then, the acquired point cloud is fitted on a spherical model and the cloud-fitted model distance is computed. Similarly, Fig. 6.17 show that the D415^{TM} produces a much smoother point cloud than the D435^{TM}, with a narrower distribution for the errors. Note the depth quantization effect on the point clouds, more evident on the D415^{TM}.

6.3.4   Mixed Pixels Error

Mixed pixels errors are present because of an average operation performed on every single pixel with its surrounding. The Kinect V2^{TM} showed to be subject to mixed pixel. We show on Fig. 6.18 that camera based on active stereoscopy are not affected by the mixed pixel effect. Nevertheless, a gap is created between depth edges, making it difficult to estimate object borders. Figure 6.18 depicts the same edge measurement with three different presets, showing the missing data.

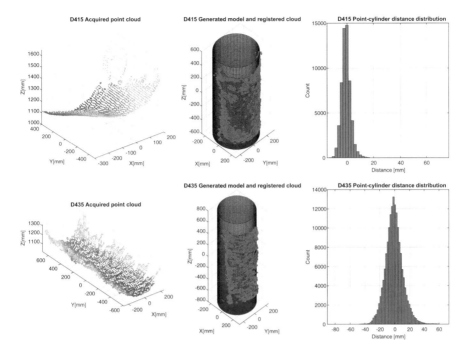

Fig. 6.16 Acquired point cloud, fitting cylinder and distance distribution. **Top:** results for the D415^{TM}. **Bottom:** results for the D435^{TM}

6.4 Conclusion

In this chapter, we provide a metrological analysis of the new generation RS400^{TM} of RealSense 3D camera provided by Intel. The RS400^{TM} generation include two main module, the D410^{TM} and the D430^{TM}, that are mounted with the D415^{TM} and the D435^{TM} devices, commercially available to the mass market. Both cameras come with different presets, that optimize the elaboration parameters of the depth map for specific applications. We show that only the default settings provides repeatable and statistically acceptable noise, that we characterized.

Analysis the uncertainty in a pixel-wise fashion, we can notice a similarity between the D415^{TM} device based on active stereoscopy and the Orbbec Astra S^{TM} device based on structured-light. The casual component of the uncertainty grows quadratically with the depth, in contrast with the linear rate of the Kinect V2^{TM} Time-of-Flight (TOF) camera. Still, such casual component is the double of the Orbbec Astra S^{TM} device uncertainty. The systematic component of the uncertainty also grows quadratically with the depth, but at a quarter of the rate of the structured-light based device, which is a considerable improvement. Nevertheless, the Kinect V2^{TM} TOF camera still provide state-of-the-art performances. We provide in Table 6.2 an overview of the depth uncertainty for the three 3D camera we present in this manuscript

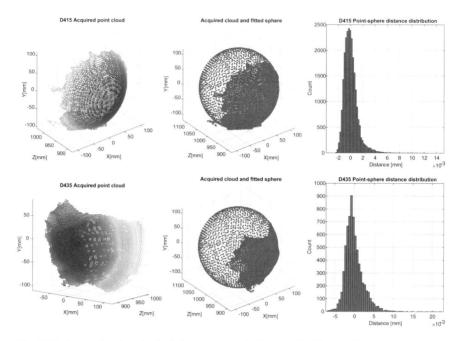

Fig. 6.17 Acquired point cloud, fitting sphere and distance distribution. **Top:** results for the D415^{TM}. **Bottom:** results for the D435^{TM}

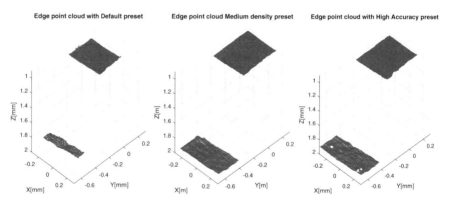

Fig. 6.18 D415^{TM} edge acquisition with "Default", "Medium density" and "High accuracy" presets

The uncertainty evaluation along the sensor plane shows the distribution of the error. In particular, we show that the projected pattern influences the quality of the depth estimation. The random component is influenced by the projected pattern, showing a regular pattern of oscillation. The systematic component is influenced by the non-ideality of the lens.

Regarding the sensor-wise analysis, we show that both camera were able to reconstruct simple geometry. However it looks like the D415^{TM} is more precise

Table 6.2 Uncertainty and bias comparison at different depth

	At 1000 mm		At 2500 mm	
	Uncertainty	Bias	Uncertainty	Bias
Kinect V2^{TM}	$\simeq$1.5 mm	$\simeq$5 mm	$\simeq$2 mm	$\simeq$10 mm
Orbbec Astra S^{TM}	$\simeq$1 mm	$\simeq$8 mm	$\simeq$5 mm	$\simeq$96 mm
D415^{TM}	$\simeq$1.5 mm	$\simeq$2 mm	$\simeq$15 mm	$\simeq$25 mm

that the D435^{TM}. We believe it is partially due to the larger FOV of the D435^{TM}, that leads into a worst spatial resolution, hence a worst precision in the triangulation process. Finally, we show that the structured-light device is not influenced by the mixed pixel error, unlike the Kinect V2^{TM}.

Overall, the SR400^{TM} generation of 3D camera provided by Intel proved to have outstanding performances when compared to other triangulation-based devices. Nevertheless, triangulation-based method do not outperform the Still, TOF device requires a powerful illumination system, definitively not energy-efficient. Hence, for embedded applications, the RS400^{TM} generation is a valuable device for shape acquisition.

Chapter 7
Conclusion

This manuscript presents the technology on which are based Three-dimension (3D) cameras. We provide a comparison of the main 3D camera available on the mass market, as well as a deep analysis of the uncertainty in the depth measurement for each technology.

First of all, we show that 3D cameras are being more and more popular in the last years. Several large tech companies such as Microsoft, Google, Apple and Intel are entering the market, especially with the plethora of applications such 3D cameras enhance. The technology is currently split between three main technologies: Time-of-Flight (TOF), Structured-Light and Active Stereoscopy. Structured-Light and Active Stereoscopy works in a similar fashion, they use a triangulation process to estimate the depth, while TOF uses a direct depth measurement.

The Microsoft Kinect V2^{TM} showed state-of-the-art performances for a 3D TOF camera. We show that the uncertainty in the depth measurement using a TOF camera scales linearly with the depth hence provide reliable measurement on longer range. The Orbbec Astra S^{TM} showed state-of-the-art performances for a 3D Structured-Light camera and The Intel RS400^{TM} generation showed state-of-the-art performances for a 3D Active Stereoscopy camera. Both Structured-Light and Active Stereoscopy are based on the same triangulation principle, which provides a depth measurement the which uncertainty grows quadratically, hence its usage is preferred for short range application.

Qualitatively, we showed that the main drawback of TOF camera reside in the multiple path effect and the mixed pixel. Both drawbacks are actually absent in the triangulation-based technology. Regarding mobile application, it is important to underline that TOF technology requires powerful LEDs, hence is not power efficient at all. Both Structured-Light and Active Stereoscopy triangulates on visual features, that just need to be visible to the cameras. On the other hand, for outdoor application, TOF provides powerful-enough signal to estimate the depth even with strong lighting

S. Giancola et al., *A Survey on 3D Cameras: Metrological Comparison of Time-of-Flight, Structured-Light and Active Stereoscopy Technologies*, SpringerBriefs in Computer Science, https://doi.org/10.1007/978-3-319-91761-0_7

condition. Both Structured-Light and Active Stereoscopy are drastically sensitive to the environmental conditions. Overall, a proper knowledge of the underlying technology will permit to take the most of it. In some devices such as the D400, predefined settings help the user to optimize the device on the specific application.

With this book, we tried to cover the technology currently available. We approach the problem from a metrological point of view, by considering the 3D cameras as instruments for depth measurements. However, estimating the depth from visual clue is a very hot topic. From a software prospective, current research makes use of deep learning method to estimate depth from a single RGB frame, provide reliable enough depth estimation. From a hardware prospective, people try to fuse several depth principle in order to cope with the drawback of each technology.

References

Cyrus S Bamji, Patrick O'Connor, Tamer Elkhatib, Swati Mehta, Barry Thompson, Lawrence A Prather, Dane Snow, Onur Can Akkaya, Andy Daniel, Andrew D Payne, et al. A 0.13 μm cmos system-on-chip for a 512×424 time-of-flight image sensor with multi-frequency photo-demodulation up to 130 mhz and 2 gs/s adc. *IEEE Journal of Solid-State Circuits*, 50(1): 303–319, 2015.

Paul J Besl and Neil D McKay. Method for registration of 3-d shapes. In *Sensor Fusion IV: Control Paradigms and Data Structures*, volume 1611, pages 586–607. International Society for Optics and Photonics, 1992.

Ayush Bhandari, Achuta Kadambi, Refael Whyte, Christopher Barsi, Micha Feigin, Adrian Dorrington, and Ramesh Raskar. Resolving multipath interference in time-of-flight imaging via modulation frequency diversity and sparse regularization. *Optics letters*, 39(6):1705–1708, 2014.

IEC BIPM, ILAC IFCC, IUPAP IUPAC, and OIML ISO. Evaluation of measurement dataguide for the expression of uncertainty in measurement. jcgm 100: 2008. 2008.

Duane C Brown. Decentering distortion of lenses. *Photometric Engineering*, 32(3):444–462, 1966.

Thomas Butkiewicz. Low-cost coastal mapping using kinect v2 time-of-flight cameras. In *Oceans-St. John's, 2014*, pages 1–9. IEEE, 2014.

Monica Carfagni, Rocco Furferi, Lapo Governi, Michaela Servi, Francesca Uccheddu, and Yary Volpe. On the performance of the intel sr300 depth camera: metrological and critical characterization. *IEEE Sensors Journal*, 17(14):4508–4519, 2017.

Edoardo Charbon, Matt Fishburn, Richard Walker, Robert K Henderson, and Cristiano Niclass. Spad-based sensors. In *TOF range-imaging cameras*, pages 11–38. Springer, 2013.

Katherine Creath. V phase-measurement interferometry techniques. *Progress in optics*, 26: 349–393, 1988.

Adrian A Dorrington, John Peter Godbaz, Michael J Cree, Andrew D Payne, and Lee V Streeter. Separating true range measurements from multi-path and scattering interference in commercial range cameras. In *IS&T/SPIE Electronic Imaging*, pages 786404–786404. International Society for Optics and Photonics, 2011.

Péter Fankhauser, Michael Bloesch, Diego Rodriguez, Ralf Kaestner, Marco Hutter, and Roland Siegwart. Kinect v2 for mobile robot navigation: Evaluation and modeling. In *Advanced Robotics (ICAR), 2015 International Conference on*, pages 388–394. IEEE, 2015.

© The Author(s), under exclusive licence to Springer International Publishing AG, part of Springer Nature 2018
S. Giancola et al., *A Survey on 3D Cameras: Metrological Comparison of Time-of-Flight, Structured-Light and Active Stereoscopy Technologies*, SpringerBriefs in Computer Science, https://doi.org/10.1007/978-3-319-91761-0

Martin A Fischler and Robert C Bolles. Random sample consensus: a paradigm for model fitting with applications to image analysis and automated cartography. In *Readings in computer vision*, pages 726–740. Elsevier, 1987.

Nicola Giaquinto, Giuseppe Maria D'Aucelli, Egidio De Benedetto, Giuseppe Cannazza, Andrea Cataldo, Emanuele Piuzzi, and Antonio Masciullo. Accuracy analysis in the estimation of tof of tdr signals. In *2015 IEEE International Instrumentation and Measurement Technology Conference (I2MTC) Proceedings*, pages 187–192. IEEE, 2015.

Nicola Giaquinto, Giuseppe Maria D'Aucelli, Egidio De Benedetto, Giuseppe Cannazza, Andrea Cataldo, Emanuele Piuzzi, and Antonio Masciullo. Criteria for automated estimation of time of flight in tdr analysis. *IEEE Transactions on Instrumentation and Measurement*, 65(5): 1215–1224, 2016.

Miles Hansard, Seungkyu Lee, Ouk Choi, and Radu Patrice Horaud. *Time-of-flight cameras: principles, methods and applications*. Springer Science & Business Media, 2012.

Richard Hartley and Andrew Zisserman. *Multiple view geometry in computer vision*. Cambridge university press, 2003.

Eli Horn and Nahum Kiryati. Toward optimal structured light patterns1. *Image and Vision Computing*, 17(2):87–97, 1999.

Texas Instruments. Introduction to the time-of-flight (tof) system design. *Users Guide*, 2014.

Leonid Keselman, John Iselin Woodfill, Anders Grunnet-Jepsen, and Achintya Bhowmik. Intel realsense stereoscopic depth cameras. *arXiv preprint arXiv:1705.05548*, 2017.

Larry Li. Time-of-flight camera – an introduction. *Texas Instruments-Technical White Paper*, 2014.

Giacomo Mainetti. Calibrazione di telecamere per sistemi di visione stereoscopica: Confronto tra algoritmi genetici e tecniche tradizionali. 2011.

Dario Piatti and Fulvio Rinaudo. Sr-4000 and camcube3. 0 time of flight (tof) cameras: Tests and comparison. *Remote Sensing*, 4(4):1069–1089, 2012.

Holger Rapp. Experimental and theoretical investigation of correlating tof-camera systems. Master's thesis, 2007.

Thikhathali Terence Ratshidaho, Jules Raymond Tapamo, Jonathan Claassens, and Natasha Govender. An investigation into trajectory estimation in underground mining environments using a time-of-flight camera and an inertial measurement unit. *South African journal of industrial engineering*, 25(1):145–161, 2014.

Michele Russo, Giorgia Morlando, and Gabriele Guidi. Low-cost characterization of 3d laser scanners. In *Videometrics IX*, volume 6491, page 649107. International Society for Optics and Photonics, 2007.

Joaquim Salvi, Jordi Pages, and Joan Batlle. Pattern codification strategies in structured light systems. *Pattern recognition*, 37(4):827–849, 2004.

Jamie Shotton, Andrew Fitzgibbon, Mat Cook, Toby Sharp, Mark Finocchio, Richard Moore, Alex Kipman, and Andrew Blake. Real-time human pose recognition in parts from single depth images. In *Computer Vision and Pattern Recognition (CVPR), 2011 IEEE Conference on*, pages 1297–1304. Ieee, 2011.

Jan Smisek, Michal Jancosek, and Tomas Pajdla. 3d with kinect. In *Consumer depth cameras for computer vision*, pages 3–25. Springer, 2013.

Piet Vuylsteke and André Oosterlinck. Range image acquisition with a single binary-encoded light pattern. *IEEE Transactions on Pattern Analysis and Machine Intelligence*, 12(2):148–164, 1990.

James C Wyant. Interferometric optical metrology-basic principles and new systems. *Laser Focus with Fiberoptic Technology*, 18(5):65–71, 1982.